W0259653

THE FLOWER BOOK

A BLOOM *for* EVERY DAY *of* THE YEAR

THE FLOWER BOOK

A BLOOM *for* EVERY DAY *of* THE YEAR

CONTENTS

INTRODUCTION 6

JANUARY 12

FEBRUARY 32

MARCH 50

APRIL 70

MAY 90

JUNE 110

JULY 130

AUGUST 150

SEPTEMBER 170

OCTOBER 190

NOVEMBER 210

DECEMBER 230

INDEX 250

PICTURE CREDITS 255

INTRODUCTION

In the kaleidoscopic world of gardening, few things evoke as much wonder and delight as the sight of a flower in full bloom. Flowers, with their myriad colours, shapes, and fragrances, have been central to human culture for thousands of years. They have adorned our gardens, embellished our celebrations, and even been pivotal in our expressions of love and sorrow.

The Flower Book, created by the team behind the BBC's *Gardeners' World Magazine*, is a comprehensive guide which pairs each day of the year with a specific flower, chosen to match the time of year it is in bloom.

The historical narratives contained within this book trace the origins and cultural significance of flowers. You'll discover the magical world of plant names – plants that were named after plant hunters, myths and folktales, their appearance or medicinal properties. Lungwort was named for its leaves, which supposedly resembled diseased lungs, while aquilegia has the common name 'Granny's bonnet' because of the shape of its flowers. You'll learn how people in France give sprigs of lily of the valley to friends and family at the *Fête du Muguet* on 1 May, Labour Day, as a symbol of good luck, or how the hawthorn was a Medieval symbol of fertility. These stories provide a deeper understanding of how flowers have been woven into the fabric of human life.

1
2
3

3
4
2
1
5
6

In addition to historical context, you will find within some expert advice on planting and growing each flower, including information on ideal soil types, height and spread, sunlight requirements, and key growing seasons. Find out when dahlias need to be lifted, which clematis flowers over winter and perennials that bloom for months on end.There are flowers to suit every garden, whether it's winter aconites to light up the ground beneath trees in winter, or a tree to attract bees and butterflies, like *Buddleia globosa* with its round orange flowers. Modern cultivars are paired with photographs that capture their vivid colours and intricate details. Older cultivars are accompanied by botanical illustrations, which offer a glimpse into the artistry and scientific inquiry of past generations.

The team at *Gardeners' World Magazine* have selected the flowers for each day based on the time of the year in which they usually bloom. Each entry provides comprehensive details on the flower's botanical characteristics, including its growth habit, bloom time, and a guide to potential pests and diseases. For instance, daffodils are known for their early spring bloom and are relatively pest-resistant, while peonies bloom later in the season and may require staking to support their heavy blossoms.

One should also look out for descriptions of the flower's fragrance, texture, and even the sounds they may produce in the wind. For instance, the intoxicating scent of jasmine or the soothing rustle of ornamental grasses can add a new dimension to the gardening experience throughout the year.

Throughout history, flowers have played a pivotal role in art and literature, as well as science and medicine. In considering a flower a day, we can understand flowers as not merely lovely things to grow in our gardens, but also as symbols of beauty and creativity, cherished by artists and gardeners alike.

We'd love you to think of this book as an opportunity to return to the tactile and immersive experience of gardening. It is a daily reminder of the simple yet profound joys of tending to a garden, observing the changing seasons, and witnessing the beauty of flowers.

Moreover, the book is a testament to the enduring appeal of gardening as both a hobby and a form of therapy. Studies have shown that gardening can reduce stress, improve your mood, and even enhance cognitive function. By engaging with the natural world, gardeners cultivate not only plants but also their own sense of well-being and connection. *The Flower Book* encourages readers to embrace these benefits, making gardening a part of their daily lives.

This book is a celebration of the botanical world. We'd like to invite you to embark on a year-long journey, discovering new flowers, rekindling old favourites, and deepening your appreciation for nature. Whether you are an experienced horticulturist, a casual gardener, or someone who simply enjoys the beauty of flowers, this book offers knowledge, inspiration, and a daily dose of floral delight.

JANUARY

1 JANUARY

Winter aconite *Eranthis hyemalis*

PLANT TYPE: Bulb/tuberous perennial | **FAMILY:** Ranunculaceae | **HEIGHT:** 10cm | **SPREAD:** 10cm | **EXPOSURE:** Full sun, Partial shade | **ASPECT:** South, north, east or west facing

Winter aconites have yellow buttercup-like flowers that appear in late winter, from January to February. This woodland plant will naturalise underneath trees or in grass, spreading to create a carpet of early spring colour. It's a good plant for bees and has been awarded an AGM by the RHS. Winter aconites grow from tubers, although they are often listed as bulbs. Tubers can be planted in the autumn, although plants 'in the green' establish better, when they are planted during or after flowering. They will thrive in humus-rich soil that doesn't dry out in the summer.

2 JANUARY

Snowdrop *Galanthus nivalis*

PLANT TYPE: Bulb | **FAMILY:** Amaryllidaceae | **HEIGHT:** 15cm | **SPREAD:** 8cm | **EXPOSURE:** Dappled shade, Partial shade | **ASPECT:** South, north, east or west facing

Snowdrops provide a cheery welcome, being one of the earliest flowers. They are very hardy, thrive in cold weather and will flower even in snow. Native to Europe, the first record of snowdrops being grown in England dates from the 16th century. With over 1,000 varieties, differing sometimes only by tiny markings on the petals, snowdrops have become a collectors' item, sought after by galanthophiles. Grow snowdrops in humus-rich soil. They establish best when planted 'in the green' in February to March. They look good in drifts, complementing other early spring flowers such as aconites and hellebores.

3 JANUARY

Sowbread *Cyclamen coum*

PLANT TYPE: Bulb | **FAMILY:** Primulaceae | **HEIGHT:** 8cm | **SPREAD:** 10cm | **EXPOSURE:** Dappled shade, Full shade, Partial shade | **ASPECT:** South, north, east or west facing

Cyclamen coum flowers between January and March, in shades from deep pink through to light pink and pure white. The petals are often described as 'upswept', curving towards the sky as if they've been blown up by the wind. The foliage is also attractive, with rounded green leaves marked by silvery lines. This hardy bulb will naturalise well around trees and shrubs in humus-rich soil. It's a good plant for shade, making a striking partner for snowdrops, ferns and aconites. Plant the corms in September or October. Alternatively, plant it while in flower or leaf and it will return the following year.

4 JANUARY

Winter heath heather *Erica carnea* 'Myretoun Ruby'

PLANT TYPE: Evergreen shrub | **FAMILY:** Ericaceae | **HEIGHT:** 15cm | **SPREAD:** 45cm | **EXPOSURE:** Full sun, Partial shade | **ASPECT:** South, north, east or west facing

Winter heather can be upright or prostrate, providing bright colour with its deep magenta-pink flowers in the colder months when little else is in bloom. 'Myretoun Ruby' is a mat-forming variety with attractive evergreen foliage, which can be used as a groundcover plant between other shrubs and perennials or as an edging plant to add contrast to upright plants in a container. It's long flowering, from January to May, and attracts pollinators with its mass of nectar-rich flowers. It grows best in neutral to acid soil, but can cope in mildly alkaline soils. Cut back after flowering to encourage good growth and tidy up the plant.

---

5 JANUARY

Pansy *Viola* × *wittrockiana* varieties

PLANT TYPE: Annual | **FAMILY:** Violaceae | **HEIGHT:** 20cm | **SPREAD:** 30cm | **EXPOSURE:** Dappled shade, Full sun | **ASPECT:** South, east or west facing

'Wittrockiana' varieties are named after the Swedish botanist Veit Brecher Wittrock, who specialised in the *Viola* genus. There are several cultivars that flower in winter, including the Frizzle Sizzle series and 'Ultima Morpho'. Winter-flowering pansies are short-lived perennials, although they are usually used as annual bedding in containers, hanging baskets and borders. They can be confused with violas, but generally violas have smaller flowers. Pansies are valuable for providing colour in winter and early spring, with a huge range of colours available including pink, purple, yellow, blue and white, as well as bi-coloured varieties. In very cold weather, pansies may stop flowering but will start again when the weather warms up. Plant in late summer to mid-autumn for winter flowers.

6 JANUARY

Mimosa *Acacia dealbata*

PLANT TYPE: Evergreen tree | **FAMILY:** Fabaceae | **HEIGHT:** 12m | **SPREAD:** 4m | **EXPOSURE:** Full sun | **ASPECT:** South or west facing

Mimosa, also known as blue wattle, is native to Australia. In the UK, it's half hardy, which means it needs a sheltered spot and is best grown in mild or coastal regions. It's grown for its bright-yellow, ball-shaped flowers, which appear from February to April. These are scented and attractive to bees. The fern-like foliage is also striking and offers interest year round. Plant at the back of a border as this tree is fast growing and can reach 12m when grown in the right conditions. Mimosa will thrive in acid to neutral soil, in a sunny spot. It only needs pruning if you want to keep it to a certain size, or to reshape it. If so, prune after flowering.

7 JANUARY

White forsythia *Abeliophyllum distichum* Roseum Group

PLANT TYPE: Deciduous shrub | **FAMILY:** Oleaceae | **HEIGHT:** 1.5m | **SPREAD:** 1.5m | **EXPOSURE:** Full sun, Light shade | **ASPECT:** South, east or west facing

The scented, star-shaped white/pale pink flowers appear on bare branches in February, followed by dark green leaves that turn purple in the autumn. It's a good choice for planting near a path, where the winter fragrance can be appreciated. Plant it in a sunny spot, in moist well-drained soil. Prune after flowering each year. It's a good plant partner for other scented, early spring-flowering shrubs such as winter honeysuckle or summer-flowering perennials and shrubs that can offer interest after the forsythia's flowers have faded.

8 JANUARY

Pansy *Viola × wittrockiana* 'Frizzle Sizzle Burgundy'

PLANT TYPE: Annual | **FAMILY:** Violaceae | **HEIGHT:** 15cm | **SPREAD:** 15cm | **EXPOSURE:** Dappled shade, Full sun | **ASPECT:** South, east or west facing

This showstopping pansy can be in bloom between October and May. It's one of the earliest to flower, with ruffled purple petals on double flowers that will add impact to any pot display. Their long flowering and hardy nature make them a popular choice for winter containers and windowboxes, combined with plants such as winter heathers, skimmia and early spring-flowering bulbs. This is a vigorous F1 variety and an excellent bedding plant. To keep them flowering, water and feed regularly, especially from early spring.

---

9 JANUARY

Oregon grape *Mahonia × media* 'Winter Sun'

PLANT TYPE: Evergreen shrub | **FAMILY:** Berberidaceae | **HEIGHT:** 4m | **SPREAD:** 2.5m | **EXPOSURE:** Partial shade, Full shade, Full sun | **ASPECT:** South, north, east or west facing

Mahonia provides bright colour at a grey time of year, with clusters of scented yellow flowers. These are a valuable source of nectar and pollen for winter-active bees and other pollinators when few plants are in flower. The flowers are followed by purple to black berries. 'Winter Sun' offers interest year-round with its spiny, evergreen foliage. It is a good focal point in a shady border and can be underplanted with a mix of complementary shade lovers such as bergenia and epimediums. Mahonia will grow best in partial shade, although it can tolerate either full sun or full shade. If grown in full sun, the soil needs to be moist.

---

10 JANUARY

Witch hazel *Hamamelis × intermedia* 'Jelena'

PLANT TYPE: Deciduous shrub | **FAMILY:** Hamamelidaceae | **HEIGHT:** 4m | **SPREAD:** 4m | **EXPOSURE:** Full Sun, Partial shade | **ASPECT:** South, north, east or west facing

The plant breeder Robert Buelder named this witch hazel after his wife 'Jelena' in the 1950s. It has curled flowers that are red at the base, lightening to yellow and orange at the tips. These appear in January and February on bare stems after the leaves have fallen in autumn. The foliage is also a high point, with vibrant autumn colour. Although the flowers have a citrussy scent, it's lighter than other witch hazel varieties and most noticeable on warm, sunny days. 'Jelena' grows best in acid to neutral, well-drained soil.

11 JANUARY

Winter honeysuckle *Lonicera fragrantissima*

PLANT TYPE: Deciduous shrub | **FAMILY:** Caprifoliaceae | **HEIGHT:** 2m | **SPREAD:** 3m | **EXPOSURE:** Dappled shade, Full sun | **ASPECT:** South or west facing

Winter honeysuckle was introduced to the UK by 19th century Scottish plant hunter Robert Fortune. The plant originates from China but is hardy in Britain, offering winter scent and white flowers between December and March. The flowers are a magnet for winter-active bumblebees and are sometimes followed by dull red berries. It produces an abundance of flowers when grown against a warm wall, but can also be grown as a free-standing shrub, mixing well with other winter-flowering plants such as hellebores and snowdrops.

12 JANUARY

Wintersweet *Chimonanthus praecox*

PLANT TYPE: Deciduous shrub | **FAMILY:** Calycanthaceae | **HEIGHT:** 4m | **SPREAD:** 2.5m | **EXPOSURE:** Full sun | **ASPECT:** South or east facing

Named wintersweet for the exceptional fragrance of its flowers, this bushy shrub is perfect for a south-facing border. It flowers from December to February, with individual yellow flowers dotted along the bare stems, rather than clusters. It's popular with florists as the stems can be cut to add colour and fragrance to flower arrangements. Wintersweet makes a good feature plant for winter, especially near a path where you can enjoy the scent, but after February it doesn't have much impact. It's worth planting it among other plants that can take over the show from March onwards.

13 JANUARY

Winter daphne *Daphne odora*

PLANT TYPE: Evergreen shrub | **FAMILY:** Thymelaeaceae | **HEIGHT:** 1.5m | **SPREAD:** 1.5m | **EXPOSURE:** Full sun, Partial shade | **ASPECT:** South or west facing

Winter daphne is native to China and Japan but is hardy throughout most of the UK. It may struggle in particularly exposed or cold locations. Daphnes are ideal for a mixed sunny border where their scented flowers will attract bees and brighten displays. *Daphne odora* blooms between December and March, with clusters of highly fragrant deep pink and white flowers. Its evergreen foliage provides year-round interest. For striking variegation, try *Daphne odora* 'Aureomarginata', which has glossy green leaves edged with yellow and purple-pink flowers with white insides.

14 JANUARY

Silk tassel bush *Garrya elliptica*

PLANT TYPE: Evergreen shrub | **FAMILY:** Garryaceae | **HEIGHT:** 4m | **SPREAD:** 4m | **EXPOSURE:** Full sun, Partial shade | **ASPECT:** South, north, east or west facing

This is a good plant for beginners as it's easy to grow and low maintenance. The hardy silver tassel bush can be grown as a wall shrub, to show off its long ornamental catkins, which grow to around 15cm. These grey-green tassels make an impressive display in January and February. The evergreen foliage looks good year round, adding privacy to boundaries or as good coverage for bare walls. Some popular cultivars include 'James Roof', which has dark green leaves and silver-grey catkins, up to 20cm long. It will grow best in a sheltered, warm spot.

---

15 JANUARY

Cornelian cherry *Cornus mas*

PLANT TYPE: Deciduous shrub | **FAMILY:** Cornaceae | **HEIGHT:** 4m | **SPREAD:** 4m | **EXPOSURE:** Full sun, Partial shade | **ASPECT:** South, north, east or west facing

Bright yellow flowers appear on bare branches from January to February, bringing welcome colour in the colder months. The flowers on this shrub or small tree are good for pollinators and will sometimes last into March. *Cornus mas* has another season of interest in the autumn, when its oval leaves turn purple. *Cornus mas* is part of the dogwood family. The fruits are edible and rich in vitamin C, with a plum-like taste when ripe. In Europe and Asia the cherry-like fruits are often used in preserves and pickles.

16 JANUARY

Blue-grey hellebore *Helleborus lividus*

PLANT TYPE: Evergreen perennial | **FAMILY:** Ranunculaceae | **HEIGHT:** 35cm | **SPREAD:** 35cm | **EXPOSURE:** Full sun, Partial shade | **ASPECT:** South or west facing

This attractive hellebore has green, cup-shaped flowers and marbled leaves. Being fairly compact it makes a good plant for a winter container, combined with other winter highlights such as snowdrops, ivy or pansies. *Helleborus lividus* is native to the Balearic Islands, such as Majorca, and some nurseries list this plant as tender, but according to the RHS it's hardy down to −10°C. This means that it should thrive in most areas of the UK, although it may struggle in cold regions if there is extreme weather. It will grow well in a sheltered spot, in moist well-drained soil.

17 JANUARY

Winter jasmine *Jasminum nudiflorum*

PLANT TYPE: Deciduous shrub | **FAMILY:** Oleaceae | **HEIGHT:** 2.5m | **SPREAD:** 2.5m | **EXPOSURE:** Full sun, Partial shade | **ASPECT:** South, north, east or west facing

Although winter jasmine is classified as a deciduous shrub, it's most often grown as a climber. It has long stems that can be tied into supports to cover a fence or wall. Its flowers are small but a vivid yellow, perfect for making an impact on a dull winter's day. Winter jasmine was introduced to the UK from China in 1844. It's hardy and easy to grow. Prune annually to keep the plant bushy and flowering well. The best time to prune is after flowering. On older plants, it is worth removing a quarter of the old shoots each year, cutting down to the base.

18 JANUARY

Hellebore *Helleborus* 'Walberton's Rosemary'

PLANT TYPE: Semi-evergreen perennial | **FAMILY:** Ranunculaceae | **HEIGHT:** 40cm | **SPREAD:**40cm | **EXPOSURE:** Full sun, Partial shade | **ASPECT:** South, north, east or west facing

Hellebore 'Walberton's Rosemary' was discovered by David Tristram of Walberton's Nursery and launched in 2009. He named the new variety after his wife, Rosemary. It's a cross between *Helleborus niger* and *H* × *hybridus* and produces a mass of pink flowers between January and April. The flowers are also upward facing, rather than drooping down. Add to woodland gardens, or plant beneath trees and shrubs for a beautiful spring display. It will thrive in partial shade in a sheltered spot, with humus-rich soil. Remove old flowers and leaves after flowering is over.

19 JANUARY

Fern-leaved clematis *Clematis cirrhosa* var. *balearica*

PLANT TYPE: Climber | **FAMILY:** Ranunculaceae | **HEIGHT:** 2.5m | **SPREAD:** 1.5m | **EXPOSURE:** Full sun | **ASPECT:** South or west facing

This winter-flowering clematis is ideal for a boundary as it has evergreen, bronze-tinted leaves. As well as providing coverage and privacy year-round its flowers that appear through December to February are a highlight for the cold months. It's a good choice for a fence or arch that is near the house so that the dainty creamy-white flowers can be appreciated. They are speckled purple inside and have a light scent. It's also a good plant for any winter-active pollinators. Fern-leaved clematis is easy to look after, requiring no regular pruning unless its size needs to be restricted.

20 JANUARY

Christmas rose *Helleborus niger*

PLANT TYPE: Semi-evergreen perennial | **FAMILY:** Ranunculaceae | **HEIGHT:** 45m | **SPREAD:** 45cm | **EXPOSURE:** Partial shade | **ASPECT:** North, east or west facing

Helleborus niger, also known as the Christmas Rose, will flourish in partially shady borders or beneath deciduous trees. It's extremely hardy and will survive temperatures down to −20°C, and is reliable with graceful white flowers that attract bees. These blooms appear between December and February, on upright stems. The leathery leaves are evergreen, which adds interest year round. Don't move hellebores once established, they will self-seed to produce new plants. Cut back old leaves and flowers as new foliage emerges.

21 JANUARY

Stinking hellebore *Helleborus foetidus*

PLANT TYPE: Evergreen perennial | **FAMILY:** Ranunculaceae | **HEIGHT:** 80cm | **SPREAD:** 45cm | **EXPOSURE:** Full sun, Partial shade | **ASPECT:** South, north, east or west facing

Named the 'stinking hellebore' because of the smell of its leaves, *Helleborus foetidus* adds architectural impact with its unusual flowers. The small, bell-shaped flowers are green with a purple edge, held on tall, upright plants up to 80cm high. It's a good choice for a shady border and makes a striking partner for plants, such as ajuga and evergreen ferns, as well as other spring flowers that thrive in dappled shade, such as muscari and dwarf narcissi. The foliage is more finely divided than other hellebores, adding another point of interest to a plant that looks good all year. The foliage complements other later-flowering plants, such as lily of the valley.

22 JANUARY

Winter heather *Erica carnea* f. *alba* 'Springwood White'

PLANT TYPE: Evergreen shrub | **FAMILY:** Ericaceae | **HEIGHT:** 20cm | **SPREAD:** 45cm | **EXPOSURE:** Full sun, Partial shade | **ASPECT:** South, north, east or west facing

Winter heathers will thrive whatever the weather and make an attractive, low-maintenance plant for the winter garden. 'Springwood White' is popular for its dainty, white, bell-shaped flowers, with chocolate anthers. It's a wonderful groundcover plant, spreading to form a mat of bright green foliage that also helps to suppress early weeds. With flowers from January to April, this is a long-flowering plant that's a valuable addition to borders with acid soil or containers filled with ericaceous compost. As well as this, it's on the RHS list of Plants for Pollinators, providing nectar and pollen for bees and other pollinating insects.

23 JANUARY

Witch hazel *Hamamelis × intermedia* 'Diane'

PLANT TYPE: Deciduous shrub | **FAMILY:** Hamamelidaceae | **HEIGHT:** 4m | **SPREAD:** 4m | **EXPOSURE:** Full sun, Partial shade | **ASPECT:** South, north, east or west facing

Hamamelis × intermedia is a cross between the Chinese witch hazel (*Hamamelis mollis*) and the Japanese witch hazel (*Hamamelis japonica*). The cultivar 'Diane', named after the breeder's daughter, has bright red, fragrant spidery flowers on leafless branches between January and March. The flowering twigs make an attractive indoor display. In autumn, the leaves turn vibrant shades of yellow and red, adding another season of interest. Plant these reliable winter-flowering shrubs in a sunny border, in well-drained soil.

---

24 JANUARY

Viola *Viola* 'Tiger Eyes'

PLANT TYPE: Evergreen perennial | **FAMILY:** Violaceae | **HEIGHT:** 20cm | **SPREAD:** 30cm | **EXPOSURE:** Full sun, Partial shade | **ASPECT:** South, east or west facing

Violas can be perennials or annuals. The variety 'Tiger Eyes' is a perennial but often grown as an annual, used in bedding schemes or for winter container displays. It's incredibly long flowering, blooming in mild spells from October through to April. Its goldenyellow flowers have striking black veins and a black blotch in the centre that makes it an eye-catching variety. Pinch off any dead flowers to encourage more to form. Although you can buy packs of viola plants, it's also possible to grow your own from seed. They will flower within 16–22 weeks.

---

25 JANUARY

Trumpet daffodil *Narcissus* 'Rijnveld's Early Sensation'

PLANT TYPE: Bulb | **FAMILY:** Amaryllidaceae | **HEIGHT:** 35cm | **SPREAD:** 20cm | **EXPOSURE:** Full sun, Partial shade | **ASPECT:** South, east or west facing

Extend your spring bulbs season with 'Rijnveld's Early Sensation', which is one of the earliest daffodils, sometimes in flower from late December and reliably putting on a show from early January. It's also long-flowering, sometimes into March. This sturdy trumpet daffodil is perfect for spring borders and can be planted alongside late-flowering varieties such as jonquilla daffodils, which will give you four months of cheery spring colour. Plant the bulbs at one and a half times their own depth between September and November.

26 JANUARY

Amaryllis *Hippeastrum* (Spider Group) *'Evergreen'*

PLANT TYPE: Bulb | **FAMILY:** Amaryllidaceae | **HEIGHT:** 50cm | **SPREAD:** 20cm | **EXPOSURE:** Bright filtered light | **ASPECT:** East or west facing windowsill

Hippeastrum are easy to grow and are often planted to add a splash of colour for Christmas. This variety makes a change from the festive red varieties, with lime green flowers that are darker towards the centre. They flower 7–10 weeks after planting and can be planted between September and January. Plant a bulb in October and November for flowers to brighten January. Place potted bulbs in a warm spot, around 21°C, until the flowers appear, then move to a cooler location (around 18°C) in bright light, so the flowers last as long as possible. Turn your plant regularly to prevent the flower stalk growing towards the light.

27 JANUARY

Sowbread *Cyclamen coum* subsp. *coum* f. *pallidum* 'Album'

PLANT TYPE: Tuberous perennial | **FAMILY:** Primulaceae | **HEIGHT:** 10cm | **SPREAD:** 10cm | **EXPOSURE:** Light shade, Partial shade | **ASPECT:** North, east or west facing

For those who appreciate the beauty of cyclamen, *Cyclamen coum* subsp. *coum* f. *pallidum* 'Album' offers something different to the usual. It has pure white petals, swept back, with a deep purple mark at the base of each petal. The flowers are held on long, slim, red stems above glossy round leaves. This variety is in flower from January to March, perfect for lighting up areas beneath trees and shrubs, or for adding to winter containers.

28 JANUARY

Pansy *Viola* × *wittrockiana* 'Mystique Blue Whiskers'

PLANT TYPE: Annual | **FAMILY:** Violaceae | **HEIGHT:** 15cm | **SPREAD:** 15cm | **EXPOSURE:** Full sun, Partial shade | **ASPECT:** South, east or west facing

There's a huge variety of pansies to choose from. 'Mystique Blue Whiskers' is an attractive light blue with a white face that named for the dark blue lines that radiate out from its central yellow blotch. These pansies flower between October and May and are usually grown as annuals for container displays or to perk up the front of flower beds. They are a good addition to window boxes, pots and hanging baskets. Their appeal is their long flowering nature and sheer range of colour combinations. The petals are edible and sometimes used in salads.

---

29 JANUARY

Amaryllis *Hippeastrum* 'Green Valley'

PLANT TYPE: Bulb | **FAMILY:** Amaryllidaceae | **HEIGHT:** 70cm | **SPREAD:** 30cm | **EXPOSURE:** Bright filtered light | **ASPECT:** East or west facing windowsill, in a sunny spot but not direct sun

Amaryllis 'Green Valley' is a stylish bulb to grow indoors for January, with pointed petals in a light green shade. If you want a flower for the New Year, plant the bulb around the end of October. It takes around eight weeks for a flower to appear. Choose a 12cm pot and half-fill the pot with compost. Sit the bulb on top and fill in around the bulb, leaving the top, pointed half exposed. Leave your bulb in a warm, dark place for a couple of weeks, then move when a shoot emerges. You'll be rewarded with a contemporary indoor display.

30 JANUARY

Madeiran squill *Scilla madeirensis*

PLANT TYPE: Bulb | **FAMILY:** Asparagaceae | **HEIGHT:** 50cm | **SPREAD:** 20cm | **EXPOSURE:** Bright, indirect light | **ASPECT:** East or west facing windowsill

This rare bulb is native to Madeira and has only become available to buy recently. Previously it has been mainly on display in botanical gardens. It's tender so needs to be grown indoors or in a greenhouse and will flower between January and February if planted between September and December. The mauve flower is 15cm long, and looks similar to a bottlebrush, on top of 60cm stalks. This exotic flower is a showstopping plant to cheer up the winter months. The pointed leaves are sometimes decorated with small purple dots. The bulbs are large, around 10cm across. Plant at two to three times the bulb's depth.

31 JANUARY

Amaryllis 'Sumatra' *Hippeastrum* (Spider Group) 'Sumatra'

PLANT TYPE: Bulb | **FAMILY:** Amaryllidaceae | **HEIGHT:** 50cm | **SPREAD:** 20cm | **EXPOSURE:** Bright indirect light | **ASPECT:** East or west facing windowsill

Amaryllis bulbs make lovely gifts or cheering plants to try growing for flowers in winter. They are easy to grow and make a striking indoor plant. 'Sumatra' is an eye-catching variety with star-shaped flowers and thin, pink petals. The flowerheads measure around 10cm across on top of stems around 50cm tall, more compact than the tallest amaryllis, which can reach more than 75cm. Bulbs should flower again the following year. After flowering, replant the bulb in a slightly larger pot with fresh compost. Cut the flower stalk down to the base. Once planted, water well.

FEBRUARY

1 FEBRUARY

Coltsfoot *Tussilago farfara*

PLANT TYPE: Perennial | **FAMILY:** Asteraceae | **HEIGHT:** 50cm | **SPREAD:** 10cm | **EXPOSURE:** Full sun | **ASPECT:** South, east or west facing

Used for hundreds of years as an ingredient in cough medicine, *Tussilago farfara* also has the common name cough wort, as the Latin verb '*tussere*' means 'to cough'. There are now safety issues around using the herb, so it's not advised to try a DIY remedy. The dandelion-like flowers provide an early source of nectar and appear before the leaves, lasting until around April. The shape of the leaves earned the plant another common name of 'ass's foot'. This plant spreads by rhizomes and can be difficult to get rid of once planted.

2 FEBRUARY

Early dwarf daffodil *Narcissus* 'February Gold'

PLANT TYPE: Bulb | **FAMILY:** Amaryllidaceae | **HEIGHT:** 30cm | **SPREAD:** 10cm | **EXPOSURE:** Full sun, Partial shade, Dappled shade | **ASPECT:** South, north, east or west facing

The botanical name *Narcissus* comes from the Greek legend of a hunter named Narcissus, who falls in love with his own reflection. In one version of this story, he's so taken with it that he doesn't eat or sleep. In the spot where he dies, a daffodil blooms. 'February Gold' is an early and award-winning dwarf daffodil with swept-back petals that naturalises well beneath deciduous trees or in lawns, and makes a pretty spring plant for a container display. Plant bulbs at twice their own depth in the autumn.

3 FEBRUARY

Hellebore *Helleborus × hybridus* Ashwood Garden hybrids

PLANT TYPE: Perennial | **FAMILY:** Ranunculaceae | **HEIGHT:** 30cm | **SPREAD:** 30cm | **EXPOSURE:** Full sun, Partial shade | **ASPECT:** South, north, east or west facing

The hellebore is a poisonous plant but also has a history of medicinal use – in the Middle Ages it's thought to have been used as a laxative. It's also associated with witchcraft. In Elizabethan folklore it was said to repel witches and was planted near doorways for this reason. Ashwood Garden hybrids are a strain of *Helleborus × hybridus*, recognised as some of the best hellebore hybrids available with many cultivars to choose from, in colours including white, pink, purple and green. The plants are reliable and healthy, flowering from January through to April.

4 FEBRUARY

Scorpion vetch *Coronilla valentina* subsp. *glauca* 'Citrina'

PLANT TYPE: Evergreen shrub | **FAMILY:** Fabaceae | **HEIGHT:** 90cm | **SPREAD:** 90cm | **EXPOSURE:** Full sun | **ASPECT:** South or west facing

Scorpion vetch is a Mediterranean plant, preferring a sunny sheltered spot in free-draining soil. The sweet scent of its small, pea-shaped lemon flowers carries better in the warmth, a lovely bonus to a bright spring day. Plant it near a path or bench where its fragrance can be appreciated. The clusters of flowers appear over a long period from late winter through to April, while the evergreen foliage adds a useful backdrop to other colourful plants throughout the year. Once established, this shrub is drought-tolerant and low-maintenance.

---

5 FEBRUARY

Dark-coloured crocus *Crocus chrysanthus* var. *fuscotinctus*

PLANT TYPE: Bulb | **FAMILY:** Iridaceae | **HEIGHT:** 10cm | **SPREAD:** 8cm | **EXPOSURE:** Full sun | **ASPECT:** South or west facing

Crocuses are an important source of early nectar and pollen for bees emerging from hibernation on warm days in early spring. Bumblebee queens sometimes sleep in the flowers overnight, before they find a nest site, although this particular crocus is tiny – perhaps too small for overnight accommodation, but still an excellent flower for bees. This species of crocus, with its deep-golden flowers, is early to flower and will naturalise well in lawns or sunny flower beds. The flowers are scented and have purple markings on the outer petals. Plant the bulbs at a depth of 10cm in September to October.

6 FEBRUARY

Sweet violet *Viola odorata*

PLANT TYPE: Perennial | **FAMILY:** Violaceae | **HEIGHT:** 10cm | **SPREAD:** 30cm | **EXPOSURE:** Full sun, Partial shade | **ASPECT:** South, north, east or west facing

Sweet violets have been used in perfumes throughout history and were used during the 19th century to scent soaps and stationery. They are most commonly purple although can also be white or lilac, with five oval petals, held above hairy heart-shaped leaves. The flowers are also edible and can be added to salads; sometimes they are candied and used on top of cakes. Sweet violets are found mainly in woodland and at the base of hedgerows, but they also make a pretty addition to a garden, providing groundcover beneath shrubs such as roses.

7 FEBRUARY

Chinese witch hazel *Loropetalum chinense* var. *rubrum* 'Fire Dance'

PLANT TYPE: Evergreen shrub | **FAMILY:** Hamamelidaceae | **HEIGHT:** 1m | **SPREAD:** 1m | **EXPOSURE:** Full sun, Partial shade | **ASPECT:** South or west facing

Originally from woodland areas of Japan, China and Burma, Chinese witch hazel is borderline hardy and thrives in a sheltered spot. In areas that get colder than -5°C, it's best grown in a pot and brought indoors over winter. The flowers are a startling pink, set against red and purple leaves. Having both showy leaves and flowers, *Loropetalum* makes a good focal point or feature plant. It's also known as Chinese fringe flower and has similar flowers to its relation witch hazel (*Hamamelis*). These flowers are slightly scented and appear from February to April.

8 FEBRUARY

Winter heather *Erica carnea* 'Vivellii'

PLANT TYPE: Evergreen shrub | **FAMILY:** Ericaceae | **HEIGHT:** 15cm | **SPREAD:** 40cm | **EXPOSURE:** Full sun, Partial shade | **ASPECT:** South, north, east or west facing

This is a prostrate heather, spreading up to 40cm and only growing up to around 15cm high. For this reason, it's an excellent groundcover plant for gardens with acid soil. Alternatively, grow it in containers of ericaceous compost where its deep purply-pink flowers will sing out from late winter to spring. *Erica carnea* grows wild in the Alps and is unaffected by winter weather, although these heathers grow best in the sun, becoming straggly in the shade. Heather is a good planting partner for early spring bulbs like daffodils, hellebores or cyclamen.

9 FEBRUARY

Forsythia *Forsythia × intermedia* 'Spectabilis'

PLANT TYPE: Deciduous shrub | **FAMILY:** Oleaceae | **HEIGHT:** 4m | **SPREAD:** 2.5m | **EXPOSURE:** Full sun, Partial shade | **ASPECT:** South, north, east or west facing

For early spring colour, it's difficult to beat this vibrant shrub, which has deep yellow flowers on bare stems until around April. 'Spectabilis' is a vigorous variety that produces a mass of flowers, followed by light-green foliage. Combining it with a compact late spring- or summer-flowering clematis will give another season of interest, as would underplanting with early spring bulbs such as snowdrops. Forsythia is hardy and easy to grow, it's not fussy about soil and will thrive as long as it gets enough sun.

10 FEBRUARY

Witch hazel *Hamamelis × intermedia* 'Strawberries and Cream'

PLANT TYPE: Deciduous shrub | **FAMILY:** Hamamelidaceae | **HEIGHT:** 3.5m | **SPREAD:** 3m | **EXPOSURE:** Full sun, Partial shade, Dappled shade | **ASPECT:** South, north, east or west facing

Witch hazel is a versatile and useful plant – for hundreds of years its branches have been used as divining rods for finding water underground. The leaves, twigs and bark are also used in medicines to treat conditions such as damaged skin and insect bites. For the garden, witch hazels bring colour and scent. 'Strawberries and Cream' is a particularly attractive variety, with spidery flowers that are yellow at the tips and a pinky red shade at the base. The flowers last until around the end of March.

11 FEBRUARY

Hazel *Corylus avellana*

PLANT TYPE: Deciduous tree | **FAMILY:** Betulaceae | **HEIGHT:** 8m | **SPREAD:** 4m | **EXPOSURE:** Full sun, Partial shade | **ASPECT:** South, east or west facing

Corylus avellana is also known as a cobnut, and it is possible to get a crop of hazelnuts if you grow one in your garden. The nuts are also eaten by squirrels and mice – dormice eat them in preparation for hibernation, as well as birds such as wood pigeons and woodpeckers. The long yellow catkins, which are the male flowers, hang on bare branches between January and March. These provide early pollen for bees. Hazel is not only useful to wildlife but can also be coppiced to encourage young stems to grow, which are often harvested to make plant supports and fencing.

12 FEBRUARY

Clematis 'Avalanche' *Clematis × cartmanii* 'Avalanche'

PLANT TYPE: Evergreen climber | **FAMILY:** Ranunculaceae | **HEIGHT:** 4m | **SPREAD:** 2m | **EXPOSURE:** Full sun, Partial shade | **ASPECT:** South, north or east facing

Unlike the *Clematis cirrhosa* varieties, 'Avalanche' will grow well in partial shade, flowering from February into April. It has bright white flowers with yellow stamens set against evergreen foliage. The flowers are up to 4cm across – the largest in this group of evergreen clematis. It looks good scrambling through shrubs or it can be trained up a fence or obelisk to bring colour to boundaries and add height. Plant it in well-drained soil with its crown at the same level as the soil.

13 FEBRUARY

Witch hazel *Hamamelis mollis* 'Brevipetala'

PLANT TYPE: Deciduous shrub | **FAMILY:** Hamamelidaceae | **HEIGHT:** 3.5m | **SPREAD:** 3.5m | **EXPOSURE:** Dappled shade, Full sun | **ASPECT:** South, east or west facing

Witch hazel has been used for centuries to treat skin ailments and it is now used in lotions, creams and gels. These attractive shrubs are a good choice for a winter border, with dazzling flowers on bare branches that will stand out even on a dull grey day. 'Brevipetala' has ochre-yellow flowers that are red at the base. These have a strong scent, which makes the stems ideal for cutting and bringing indoors for a winter display. This witch hazel has more than one season of interest as its dark-green foliage turns a vibrant yellow and orange before falling in the autumn.

14 FEBRUARY

Persian ironwood *Parrotia persica*

PLANT TYPE: Deciduous tree | **FAMILY:** Hamamelidaceae | **HEIGHT:** 8m | **SPREAD:** 10m | **EXPOSURE:** Full Sun, Partial shade | **ASPECT:** South, east or west facing

Parrotia persica can be grown as a shrub or a tree. If left to grow as a tree, it needs a lot of space, as it can reach up to 8m tall over time. As it has multiple stems it grows well as a shrub. The red, spider-like flowers appear December to March – these don't have petals, instead their buds open to display striking red stamens. This tree also looks attractive in autumn when the leaves turn a spectrum of vibrant shades from yellow to red and purple. The bark, especially on mature trees, is attractive, turning brown and flaking.

15 FEBRUARY

Paperbush *Edgeworthia chrysantha*

PLANT TYPE: Deciduous shrub | **FAMILY:** Thymelaeaceae | **HEIGHT:** 1.5m | **SPREAD:** 1.5m | **EXPOSURE:** Full sun, Partial shade | **ASPECT:** South, east or west facing

Native to China and found in the Himalayas, paperbush is an unusual, pretty shrub that's related to the daphne, with flowers at the ends of its branches. The yellow flowers emerge from white, hairy, silky buds up until April, attracting insects with their scent. Plant it in a sheltered spot. Although it's hardy through most of the UK, it will grow best through the winter if it's not exposed to extreme cold. It isn't affected by any particular pests and diseases, making it a low-maintenance option for a winter border.

16 FEBRUARY

Early squill *Scilla mischtschenkoana*

PLANT TYPE: Bulb | **FAMILY:** Asparagaceae | **HEIGHT:** 15cm | **SPREAD:** 15cm | **EXPOSURE:** Full sun, Partial shade | **ASPECT:** South, north, east or west facing

Early squill is a good plant partner for other early spring flowers that light up the ground, such as crocus and eranthis. It would also combine nicely underneath trees with hellebores as the flowers will remain well into March. Squill are not native to the UK, originating from the mountains in Northern Iran, but are hardy enough to naturalise well in Britain, in grass or gravel, returning every year. The flowers are a pale blue and the bulbs would also be a good choice for containers, either as a layer above tulips or in a shallow terracotta pot which would show off their graceful flowers.

17 FEBRUARY

Lesser celandine *Ficaria verna* subsp. *verna*

PLANT TYPE: Perennial | **FAMILY:** Ranunculaceae | **HEIGHT:** 5cm | **SPREAD:** 15cm | **EXPOSURE:** Dappled shade, Full shade | **ASPECT:** North, east or west facing

Queen bumblebees emerging from hibernation and other early pollinators will benefit from the nectar of lesser celandine, which has small, yellow star-shaped flowers. It's a British native and belongs to the buttercup family. Sometimes considered a weed, it spreads by underground tubers. In the right place, it creates attractive groundcover and can be a useful plant to grow on a bank or below a hedge to add a boost of colour. It is mostly found in areas where there is damp shade, beneath hedges and beside streams or in woodland.

---

18 FEBRUARY

February daphne *Daphne mezereum*

PLANT TYPE: Deciduous shrub | **FAMILY:** Thymelaeaceae | **HEIGHT:** 1.5m | **SPREAD:** 1.5m | **EXPOSURE:** Full sun, Partial shade | **ASPECT:** South, east or west facing

Daphne mezereum is sometimes called the February daphne because of its early flowering nature. This makes it a valuable shrub for the border when much of the early colour comes from spring bulbs. It has exceptional scent with clusters of purple to red flowers on bare branches, which provide nectar for bees. The branches with their mass of fragrant flowers are often used for vases indoors but the scent is equally welcome when the shrub is planted near a doorway or path. The flowers fade as the leaves appear. Despite its pretty appearance, all parts of *Daphne mezereum* are toxic, especially the berries.

---

19 FEBRUARY

Algerian iris *Iris unguicularis*

PLANT TYPE: Bulb | **FAMILY:** Iridaceae | **HEIGHT:** 30cm | **SPREAD:** 30cm | **EXPOSURE:** Full sun | **ASPECT:** South, east or west facing

These iris are vigorous plants with scented purple flowers between January and March. They make attractive perennials for a late-winter to early-spring container, or to bring colour to a border, mixed in with evergreen grasses or alongside snowdrops and narcissus. The flowers have a yellow marking and white stripes near the base of the petal. This bulb is native to Africa, and ideally needs a warm place in the garden, with well-draining soil. Popular varieties include 'Mary Barnard', which has rich purple flowers or 'Walter Butt' for a pale lilac flower.

---

20 FEBRUARY

Oregon grape *Mahonia × media*

PLANT TYPE: Evergreen shrub | **FAMILY:** Berberidaceae | **HEIGHT:** 4m | **SPREAD:** 4m | **EXPOSURE:** Full sun, Partial shade | **ASPECT:** South, north, east or west facing

The allure of a mahonia lies in the dazzling yellow colour that can brighten any winter's day. The flowers are an intense shade, and scented too, providing an invaluable source of pollen and nectar for winter colonies of bumblebees and other pollinators. The flowers appear through winter into spring. It's a good plant for the back of the border, reaching up to 4m high. Although the flowers are a winter highlight, the leaves of a mahonia provide architectural interest all year round. The cultivar 'Charity' is one of the best known, with an upright habit and long leaves.

21 FEBRUARY

Common name *Crocus × cultorum* 'Jeanne D'Arc'

PLANT TYPE: Bulb | **FAMILY:** Iridaceae | **HEIGHT:** 10cm | **SPREAD:** 5cm | **EXPOSURE:** Full sun | **ASPECT:** South or west facing

In the language of flowers, where flower names are given meanings, crocus is associated with cheerfulness. This is perhaps because the sight of a carpet of yellow or purple flowers spread beneath a tree or lawn in late winter is cheering. Crocuses will bloom even through snow. Native to Eastern Europe, crocuses are a much-needed source of nectar and pollen for queen bumblebees emerging from hibernation. This variety is from a group of Dutch hybrids with larger corms. These can be planted deeper than normal crocuses. It has white goblet-shaped flowers with purple staining.

22 FEBRUARY

Iris 'Katharine Hodgkin' *Iris reticulata* 'Katharine Hodgkin'

PLANT TYPE: Bulb | **FAMILY:** Iridaceae | **HEIGHT:** 15cm | **SPREAD:** 10cm | **EXPOSURE:** Full sun | **ASPECT:** South, north, east or west facing

The name iris is associated with the Greek goddess of rainbows, who was a messenger to the gods. This iris is colourful but has the watery look of a rainbow. It is subtle, not showy, with pale-blue petals that are yellow at the base and spotted with green. Plant it in containers to appreciate the markings close up. It may look delicate but this is a tough plant that will appear whatever the weather in January and February. Appearing at the same time as crocuses, it will also grow well alongside them in borders and beneath deciduous trees.

23 FEBRUARY

Iris 'George' *Iris reticulata* 'George'

PLANT TYPE: Bulb | **FAMILY:** Iridaceae | **HEIGHT:** 12cm | **SPREAD:** 10cm | **EXPOSURE:** Full sun | **ASPECT:** South, north, east or west facing

The miniature iris variety 'George' is a more vibrant colour than 'Katharine Hodgkin', with deep purple petals, which are marked with bright yellow at the base. The flowers are scented, which is especially noticeable in the spring sunshine. Reticulata irises are a group of small, early-flowering iris species with short flower stems that come from dry areas in central Europe and central Asia. They are hardy and grow well in small pots, windowboxes, rock gardens and raised beds. If you forget to plant bulbs in the autumn, they are available in flower in the winter.

24 FEBRUARY

Danford iris *Iris danfordiae*

PLANT TYPE: Bulb | **FAMILY:** Iridaceae | **HEIGHT:** 15cm | **SPREAD:** 10cm | **EXPOSURE:** Full sun, Partial shade | **ASPECT:** South or east facing

Iris danfordiae is a reticulate iris, but a species rather than a variety – it should also be planted a little deeper than the other bulbs. There is nothing subtle about this species, which has bright yellow petals, perfect for brightening a spring day. It can be in flower between January and March. Look closely and you'll see tiny brown or black spots on the petals. This species was introduced from Turkey by Mrs C.G. Danford in 1876, who was an English plant hunter.

25 FEBRUARY

Winter heather *Erica × veitchii* 'Exeter'

PLANT TYPE: Evergreen shrub | **FAMILY:** Ericaceae | **HEIGHT:** 1.8m | **SPREAD:** 1.8m | **EXPOSURE:** Full sun | **ASPECT:** South, north, east or west facing

Heather is an excellent choice for coastal gardens as it can tolerate salt spray. This variety is a bushy shrub, reaching up to almost 2m in height, with white scented flowers. *Erica × veitchii* is a tree or shrub heather, a hybrid of the *Erica arborea* (the tree heath) and *Erica lusitanica* (the Portuguese tree heather). It'll grow in most alkaline soils and flowers from January to April, providing nectar for bees. This type of heather is popular for winter colour but also for its evergreen foliage from the deep green needles that make a good backdrop for other plants when the flowers are over.

26 FEBRUARY

Winter heather *Erica × darleyensis* 'Darley Dale'

PLANT TYPE: Evergreen shrub | **FAMILY:** Ericaceae | **HEIGHT:** 30cm | **SPREAD:** 55cm | **EXPOSURE:** Full sun | **ASPECT:** South, north, east or west facing

Known also as 'Pink Perfection', this winter heather is a versatile variety that will thrive in almost any type of soil, which is useful as many heathers need acid soil. The pretty pink flowers are nectar rich, attracting bees and pollinating insects. As with many other heathers, 'Darley Dale' is a good choice for groundcover and can be used in borders alongside hellebores and snowdrops, or beneath eye-catching shrubs like dogwood, which have vibrant stems but don't block the sunlight. It flowers through February and March.

27 FEBRUARY

Small-leaved kowhai *Sophora microphylla*

PLANT TYPE: Evergreen tree | **FAMILY:** Fabaceae | **HEIGHT:** 8m | **SPREAD:** 8m | **EXPOSURE:** Full sun | **ASPECT:** South or west facing

This is a graceful spreading shrub or small tree with dark green evergreen foliage. It's native to the Himalayas and Western China, but grows well in Britain in a sheltered spot in full sun. Each leaf has around 40 pairs of oval leaflets and the dark-yellow, pea-shaped flowers hang in clusters. One of the most popular cultivars is 'Sun King', which is bushier than the species, but not as tall, growing to around 3m. It starts flowering in February and can go on into April or May. It can be grown as a wall shrub, a stand-alone shrub or in a container.

---

28 FEBRUARY

Lungwort *Pulmonaria saccharata*

PLANT TYPE: Perennial | **FAMILY:** Boraginaceae | **HEIGHT:** 30cm | **SPREAD:** 30cm | **EXPOSURE:** Dappled shade, Partial shade, Full shade | **ASPECT:** North, east or west facing

Lungwort is named for its leaves, which are mottled and supposed to resemble a diseased lung. This observation, according to folk history, was made by early herbalists, who were looking for remedies and searched for connections between body parts and a similarity in plants. This way of using a plant's appearance to guess its medical properties was known as the doctrine of signatures. Pulmonaria is an important plant for bees, especially the hairy-footed flower bee, *Anthophora plumipes*. Pulmonarias can have flowers that are pink, purple, red, white or puple like *Pulmonaria saccharata*, which also has attractive variegated foliage.

MARCH

1 MARCH

Grape hyacinth *Muscari latifolium*

PLANT TYPE: Bulb | **FAMILY:** Asparagaceae | **HEIGHT:** 15cm | **SPREAD:** 10cm | **EXPOSURE:** Full sun, Partial shade | **ASPECT:** South, north, east or west facing

Muscari is said to have the common name grape hyacinth due to the flowers resembling upside-down bunches of grapes. This species has particularly striking, two-tone flowers, light blue on the top and dark on the bottom and slightly more like a crazy haircut on top than its neater relation, *Muscari armeniacum*. Although it looks exotic, it's just as hardy as other muscari or spring bulbs and suits being planted in big groups, either in the ground or in containers. Plant the bulbs 10cm deep in the autumn.

2 MARCH

Daffodil *Narcissus* 'Tête-à-Tête'

PLANT TYPE: Bulb | **FAMILY:** Amaryllidaceae | **HEIGHT:** 15cm | **SPREAD:** 5cm | **EXPOSURE:** Full sun, Partial shade | **ASPECT:** South, east or west facing

Each stem of this dainty daffodil has up to three flowers. Its name 'Tête-à-Tête' is said to be because the flowers appear to be talking to each other. One of the most popular dwarf varieties, it's perfect for planting in raised beds, at the front of borders or in patio containers. Combined with ivy, hellebores or muscari it's an attractive choice for an early spring window box. It can be planted in the autumn, but is often available in small pots in the spring at the garden centre, and once planted will come back every year, flowering in March and April.

3 MARCH

Siberian squill *Scilla siberica*

PLANT TYPE: Bulb | **FAMILY:** Asparagaceae | **HEIGHT:** 20cm | **SPREAD:** 5cm | **EXPOSURE:** Full sun, Partial shade | **ASPECT:** South or east facing

The Siberian squill, as might be guessed from its name, is native to Russia and the Caucasus region. This means it's incredibly hardy, growing vigorously in lawns as well as beneath shrubs and trees. It has elegant spikes of violet blue bell-shaped flowers, up to five on each stem. The flowers appear in March and April, and naturalise well. For a natural look, plant them in drifts, scattering the bulbs on the ground and planting them where they fall. They will self-seed and soon form large clumps, bringing welcome spring colour to the garden.

---

4 MARCH

Radde's fritillary *Fritillaria reddeana*

PLANT TYPE: Bulb | **FAMILY:** Liliaceae | **HEIGHT:** 60cm | **SPREAD:** 30cm | **EXPOSURE:** Full sun, Partial shade | **ASPECT:** South, east or west facing

An unusual dwarf crown imperial that has cream flowers. It's far shorter than other crown imperials at 60cm high – whereas *Fritillaria imperialis* can reach 1.2m. The colour deepens to more of a lime yellow as the flower matures. This crown imperial is native to Iran and Turkmenistan. It's extremely hardy and will grow in a range of soils in either full sun or light shade. The bulbs should be planted at four times their depth in the autumn.

---

5 MARCH

Dog's tooth violet *Erythronium dens-canis*

PLANT TYPE: Bulb | **FAMILY:** Lilaceae | **HEIGHT:** 25cm | **SPREAD:** 15cm | **EXPOSURE:** Partial shade | **ASPECT:** South, north, east or west facing

The English name for this dainty flower is associated with the shape of its bulb, which resembles a dog's tooth. *Erythonium dens-canis* is a woodland-loving plant that thrives in moist, well-drained soil in partial shade. It's perfect for the front of shady borders or planting underneath trees, where its graceful flowers will light up the ground between March and April. The flowers have violet-pink petals that curve upwards, held on slender stems above attractive foliage. The green leaves are marbled with brown. Plant the bulbs as soon as possible after receiving or buying them as they will not grow as well if they dry out.

6 MARCH

Chinese redbud *Cercis chinensis* 'Avondale'

PLANT TYPE: Tree | **FAMILY:** Fabaceae | **HEIGHT:** 4m | **SPREAD:** 2.5m | **EXPOSURE:** Full sun, Partial shade | **ASPECT:** South or west facing

This Chinese redbud cultivar is a good tree for a small garden that can be grown in a large pot for a few years. It can eventually grow to 4m tall but grows at an average rate so will remain compact for many years. The pea-like flowers are a startling pink from March to April, emerging before the leaves on bare, upright branches. The leaves also provide impact during autumn when they turn yellow. It's evident from the name that it comes from China, but it grows well in the British climate – it's fully hardy, although it may need protection for the first couple of years.

7 MARCH

Glory of the snow *Chionodoxa luciliae*

PLANT TYPE: Bulb | **FAMILY:** Asparagaceae | **HEIGHT:** 15cm | **SPREAD:** 10cm | **EXPOSURE:** Dappled shade, Full sun | **ASPECT:** South, north, east or west facing

'*Chion*' is the Greek word for snow; '*doxa*' means glory, which is where we get 'Glory of the Snow' as a name for this attractive plant. It's likely that it was named for its beauty, flowering even through a layer of snow. *Chionodoxa luciliae* comes from Eastern Europe, where it flowers at high altitudes. It has star-shaped blue flowers that look spectacular when planted in large drifts, with a display lasting for three to four weeks from February to March. This makes it a good plant partner for bulbs such as miniature irises and daffodils. Plant the bulbs in autumn, around 5cm deep in well-drained soil.

8 MARCH

Heart's ease *Viola tricolor*

PLANT TYPE: Annual or short-lived perennial | **FAMILY:** Violaceae | **HEIGHT:** 12cm | **SPREAD:** 10cm | **EXPOSURE:** Full sun, Partial shade | **ASPECT:** South, east or west facing

Heart's ease has many other common names, including 'kiss me at the garden gate', 'cat's faces' and 'Cupid's flower'. The sheer number of them suggests that this flower has been useful over a long period in herbal medicine and it's said it was used as a love potion in Medieval times. Under one of its common names, 'love in idleness', it played a part in *A Midsummer Night's Dream* as a love charm. *Viola tricolor* has purple, cream and yellow flowers and is often seen flowering in the countryside. It can be in bloom from March through to October, depending on when it's sown, and it self-seeds freely.

---

9 MARCH

Wood anemone *Anemone nemorosa*

PLANT TYPE: Bulb | **FAMILY:** Ranunculaceae | **HEIGHT:** 15cm | **SPREAD:** 15cm | **EXPOSURE:** Dappled shade, Partial shade | **ASPECT:** South, north, east or west facing

If you spot wood anemone on a walk through the woods, it's likely that you are in ancient woodland that has been around since the 1600s. The Woodland Trust use this plant and others such as bluebells as clues to the age of a wood. You can also grow these easily in your own garden, planting the rhizomes (sometimes referred to as bulbs) in September or October in moist soil. The species has white flowers in March, often flowering on into April and May. Wood anemones are poisonous so it's best to avoid handling them – the toxins can irritate skin and eyes.

10 MARCH

Red lungwort *Pulmonaria rubra*

PLANT TYPE: Deciduous perennial | **FAMILY:** Boraginaceae | **HEIGHT:** 40cm | **SPREAD:** 90cm | **EXPOSURE:** Full shade, Partial shade | **ASPECT:** North, east or west facing

Red lungwort is among one of the first pulmonarias to flower in the spring and, unusually for a shade-loving plant, it has red/coral-coloured flowers. It's also one of the first perennials to flower in the spring, mixing well with hellebores – another early-flowering perennial – as well as shade lovers such as heucheras, ferns and hostas. Plant it in moist soil where possible, avoiding areas that become dry during summer. It will flower from March to April and is known for attracting bees.

11 MARCH

Creeping forget-me-not *Omphalodes verna*

PLANT TYPE: Perennial | **FAMILY:** Boraginaceae | **HEIGHT:** 20cm | **SPREAD:** 20cm | **EXPOSURE:** Partial shade, Full shade | **ASPECT:** North, east or west facing

Omphalodes verna is a pretty groundcover plant that thrives in shade, forming a mat of blue flowers through March and April. The flowers are small and rise above heart-shaped leaves. This is the perfect plant for creating interest beneath trees and shrubs, especially in those gloomy shady areas that can be difficult to plant up. It will die back through autumn and regrow in the spring. Plant it alongside other shade lovers such as ferns or pulmonaria for a spectacular spring display.

12 MARCH

Corsican hellebore *Helleborus argutifolius*

PLANT TYPE: Perennial | **FAMILY:** Ranunculaceae | **HEIGHT:** 90cm | **SPREAD:** 50cm | **EXPOSURE:** Full sun, Partial shade | **ASPECT:** South, north, east or west facing

The Corsican hellebore is also known as the holly-leaved hellebore because of its prickly-edged leaves. In Corsica it grows in the wild and at 60cm it's one of the tallest hellebores, with flowers on tall stems. The foliage is evergreen, creating architectural impact year-round. It forms an attractive clump, making a good plant partner for other late-winter highlights such as snowdrops and epimediums. The unusual, green bowl-shaped flowers appear during the main flowering period from January to March but can be in flower until May. This is a useful plant for suppressing weeds.

13 MARCH

Native primrose *Primula vulgaris*

PLANT TYPE: Perennial | **FAMILY:** Primulaceae | **HEIGHT:** 10cm | **SPREAD:** 35cm | **EXPOSURE:** Full sun, Partial shade | **ASPECT:** South, north, east or west facing

In Irish folklore, primroses were a symbol of safety, used in doorways to protect homes from fairies. Primroses are a sign of spring, a welcome sight as we emerge from winter. They are also an early source of nectar and pollen for bees and other pollinators, such as the brimstone and small tortoiseshell butterflies, as well as a caterpillar foodplant for several species of moth. Primroses are easy to grow, thriving in damp shade, beneath trees or in woodland areas, where they will self-seed and naturalise to create increasingly colourful spring displays. The flowers are edible and have a sweet flavour, often used on top of salads.

---

14 MARCH

Elephant's ears *Bergenia* 'Sunningdale'

PLANT TYPE: Perennial | **FAMILY:** Saxifragaceae | **HEIGHT:** 45cm | **SPREAD:** 45cm | **EXPOSURE:** Full sun, Partial shade | **ASPECT:** South, north, east or west facing

Bergenia are prized not only for their upright pink flowers but their large evergreen leaves, which create excellent groundcover through the year. The size of the leaves has led to the common name of elephant's ears, and 'Sunningdale' has large green leaves that turn red through autumn and winter. Its bell-shaped flowers are pink, held on short stems in the spring. Although renowned for being a good plant for shade, bergenias are versatile and will grow well in the sun, where their leaves will take on more vibrant colour. These plants are low maintenance and drought-resistant, thriving in a variety of soils.

---

15 MARCH

Rue anemone *Thalictrum thalictroides*

PLANT TYPE: Perennial | **FAMILY:** Ranunculaceae | **HEIGHT:** 10cm | **SPREAD:** 40cm | **EXPOSURE:** Partial shade | **ASPECT:** North or east facing

Rue anemone is still sometimes listed as *Anemonella thalictroides* – it has white flowers similar to a wood anemone, and the leaves are divided and fern-like, like those of *Thalictrum*. It's a woodland plant, native to eastern parts of the USA, thriving in moist conditions in partial shade, spreading to form clumps of pretty flowers that appear until May or June. The white flowers are sometimes tinted with pink. Its spreading nature makes it a good choice for groundcover, where shrubs or trees will provide some shade.

16 MARCH

Spiny pin cushion cactus *Mammillaria spinosissima*

PLANT TYPE: Cactus | **FAMILY:** Cactaceae | **HEIGHT:** 30cm | **SPREAD:** 8cm | **EXPOSURE:** Full sun | **ASPECT:** South, north or east facing

The spiny pin cushion cactus is aptly named, with a mass of rusty-brown or white spines. It's an easy houseplant for a beginner as it's not only low-maintenance but has pretty bright pink flowers in the spring. These are funnel-shaped and there may be a few blooming at once. This genus is native to desert regions of central America, in particular Mexico. Cactus don't need watering over winter. Water only between mid-spring and summer, when the compost is dry, and avoid overwatering or the cactus will rot. Feed once a month with a special cactus fertiliser.

17 MARCH

Forget me not *Myosotis sylvatica*

PLANT TYPE: Perennial | **FAMILY:** Boraginaceae | **HEIGHT:** 30cm | **SPREAD:** 15cm | **EXPOSURE:** Full sun, Partial shade | **ASPECT:** South, north, east or west facing

Forget me nots were a Medieval symbol of love, associated with a German legend about a young man; in some stories it's a knight, who falls into the a river while picking the flowers for his fiancé. As he is swept away, he throws them to her, calling 'Do not forget me'. Forget me nots create a spread of sky-blue flowers in spring that produce a spectacular display alongside wallflowers and tulips. They grow well in moist, well-drained soil in a sunny or partially shaded spot. Leave them to self-seed or sow seed in summer for flowers the following year.

18 MARCH

Hoop petticoat daffodil *Narcissus bulbocodium*

PLANT TYPE: Bulb | **FAMILY:** Amaryllidaceae | **HEIGHT:** 15cm | **SPREAD:** 10cm | **EXPOSURE:** Full sun, Partial shade | **ASPECT:** South, north, east or west facing

This dwarf narcissi has a large, flared cup but tiny petals, hence the name Hoop petticoat daffodil. It's an unusual variety that deserves a place where its flowers can be admired, in a rockery, container or at the front of a border. It can be grown in lawns, but the grass needs to be short for it to naturalise well. Unlike other daffodils, its foliage is more like grass than leaves. There are also a white and pale lemon varieties of hoop petticoat daffodils such as 'Arctic Bells' or *Narcissus bulbocodium* var. *citrinus*.

---

19 MARCH

Early stachyurus *Stachyurus praecox*

PLANT TYPE: Deciduous tree | **FAMILY:** Stachyuraceae | **HEIGHT:** 4m | **SPREAD:** 2.5m | **EXPOSURE:** Full sun, Partial shade | **ASPECT:** South, east or west facing

Stachyurus praecox is native to Japan, growing on the edges of forests in warm parts of the country, where it thrives in moist, well-drained soil. The tiny, yellow, cup-shaped flowers appear on spikes that look a bit like catkins (racemes) dangling all along the bare branches, from March until May. This attractive shrub was introduced to the west by German explorer Philipp Franz von Siebold, who describes it in his book *Flora Japonica* in 1836, which he wrote with German physician Joseph Zuccarini. It can be grown either against a wall or in a border with other early-flowering shrubs.

20 MARCH

Greater snowdrop *Galanthus elwesii*

PLANT TYPE: Bulb | **FAMILY:** Amaryllidaceae | **HEIGHT:** 30cm | **SPREAD:** 10cm | **EXPOSURE:** Partial shade | **ASPECT:** South, north, east or west facing

Snowdrops usually appear as winter approaches its end, but this snowdrop is later flowering than most, appearing in March. Known as the greater snowdrop, it has large flowers with a green patch at the base. It's named after Henry John Elwes, an English plant collector who discovered the original plants in Turkey in 1874. He was born at Colesbourne Park, where there is now a collection of more than 350 cultivars of snowdrop. *Galanthus elwesii* grows best in heavy soil in partial shade. Plant 'in the green', rather than as bulbs, as they establish better this way.

21 MARCH

Perennial wallflower *Erysimum* 'Apricot Delight'

PLANT TYPE: Perennial | **FAMILY:** Brassicaceae | **HEIGHT:** 50cm | **SPREAD:** 50cm | **EXPOSURE:** Full sun, Partial shade | **ASPECT:** South, east or west facing

This variety is a perennial wallflower and although short-lived, it's a good investment if you like bedding wallflowers as it flowers for months but will come back the next year. It's easy to take cuttings to raise new plants, which can replace the parent plant when it dies. Its warm-orange flowers are ideal for a sunny border with a hot colour scheme. This small evergreen shrub will flower from March until October, sometimes longer into the winter months. As well as having attractive scented flowers, it's beneficial to wildlife, with nectar- and pollen-rich flowers that attract bees, butterflies and moths. Deadhead regularly to extend the flowering season and prune lightly after flowering to prevent the plant becoming leggy.

22 MARCH

Wild cherry *Prunus avium*

PLANT TYPE: Deciduous tree | **FAMILY:** Rosaceae | **HEIGHT:** 20m | **SPREAD:** 8m | **EXPOSURE:** Full sun | **ASPECT:** South, north, east or west facing

The '*avium*' part of the wild cherry's name is a reference to birds, as they play a vital role in dispersing the tree's seeds. Birds that eat the cherries include blackbird and song thrush, but the fruits are also popular with wood mice, dormice and badgers. This pretty native tree is a popular choice for large gardens, as it produces a mass of white blossom in the spring, and its dark-green leaves provide striking autumn colour, turning a bright red. The wild cherry has a wide range of cultivars to choose from including 'Plena', which has clusters of double white flowers, but no fruit.

23 MARCH

Star magnolia *Magnolia stellata*

PLANT TYPE: Deciduous shrub | **FAMILY:** Magnoliceae | **HEIGHT:** 3m | **SPREAD:** 4m | **EXPOSURE:** Full sun, Partial shade | **ASPECT:** South, east or west facing

Named the star magnolia for the shape of its flowers, *Magnolia stellata* is native to Japan, and is a beautiful shrub for a small garden. It grows to 3m but it is slow-growing and compact, with a mass of blooms that appear on bare branches. These are sometimes flushed with a hint of pink. It makes a spring highlight for a lawn or a sheltered border. *Magnolia stellata* is a low-maintenance shrub; it doesn't need regular pruning, but if there are any diseased, broken or crossing branches, these can be removed in summer. Give the tree a mulch in spring with manure and leaf mould.

---

24 MARCH

Giant wake robin *Trillium chloropetalum*

PLANT TYPE: Perennial | **FAMILY:** Melanthiaceae | **HEIGHT:** 50cm | **SPREAD:** 50cm | **EXPOSURE:** Full shade, Partial shade | **ASPECT:** North, east or west facing

The '*tri*' in *trillium* is likely to refer to the three leaves of this plant, and its flowers, which have three petals. It's said that the trio is linked to the three phases of a woman's life – from maiden to mother to old age – and the root has been used by Native American tribes to help with childbirth. The upright burgundy flowers last well into May, while the foliage looks attractive for months after flowering. The leaves are a lush green mottled with patches of brown and dark green. Trilliums need a shady spot in moist, well-drained soil, and they make good partners for other plants that enjoy these conditions, such as erythroniums.

---

25 MARCH

Marsh marigold *Caltha palustris*

PLANT TYPE: Marginal pond plant | **FAMILY:** Ranunculaceae | **HEIGHT:** 25cm | **SPREAD:** 45cm | **EXPOSURE:** Dappled shade, Full sun | **ASPECT:** South or west facing

Marsh marigolds are found in the wild growing alongside streams and in the shallow water of ponds. They are striking plants for bog gardens and the edges of ponds, forming clumps of round, scalloped leaves up to 10cm across. The large, buttercup-like flowers appear in the spring, attracting a wide variety of pollinators, such as bees. The flowers have also led to a common name of kingcup, apparently due to the fact that the flowers look like the cups of kings.

26 MARCH

Crab apple *Malus sylvestris*

PLANT TYPE: Deciduous tree | **FAMILY:** Rosaceae | **HEIGHT:** 10m | **SPREAD:** 8m | **EXPOSURE:** Full sun, Partial shade | **ASPECT:** South, north, east or west facing

Mature crab apple trees can become quite twisted and gnarled, with twigs developing spines, and this 'crab-like' appearance may have led to its common name of crab apple. Prized for their white to pink spring blossom, crab apples and its cultivars are easy to grow, producing small, green apple-like fruits that can be used in jellies. The fruit is also a source of pectin, which is used for setting jams. This attractive tree supports a variety of wildlife, from bees and other pollinators to the caterpillars of several moths. Its fruits are eaten by birds and small mammals. It grows best in moist, well-drained soil in full sun.

27 MARCH

Silver birch *Betula pendula*

PLANT TYPE: Deciduous tree | **FAMILY:** Betulaceae | **HEIGHT:** 25m | **SPREAD:** 10m | **EXPOSURE:** Full sun, Dappled shade | **ASPECT:** South, north, east or west facing

The silver or common birch is a fast-growing tree (around 50cm a year), and is popular for its silver bark, yellow autumn colour and yellow-brown spring catkins. Its narrow shape makes it an elegant addition to gardens, and its small leaves create a light canopy. It also looks good in winter, due to the white bark, and combines well with woodland plants such as hellebores. For a smaller garden, the columnar and more compact *Betula pendula* 'Fastigiata' is a good choice. Silver birch supports over 300 species of insect and attracts birds such as siskins and greenfinches, which eat the seeds.

28 MARCH

Giant golden saxifrage *Chrysosplenium macrophyllum*

PLANT TYPE: Evergreen perennial | **FAMILY:** Saxifragaceae | **HEIGHT:** 20cm | **SPREAD:** 20cm | **EXPOSURE:** Full shade, Partial shade | **ASPECT:** North or east facing

This woodland plant from China has a spreading form and white-green flowers that have pink stamens, perfect for lightening a shady border in the spring. The flowers appear above rosettes of evergreen foliage that has red tints. It's in the same family as bergenias and there are some similiarities, with the rounded leathery leaves and flowers also blooming on stalks. The plant spreads by sending out runners, creating attractive groundcover over time.

---

29 MARCH

Summer snowflake *Leucojum aestivum*

PLANT TYPE: Bulb | **FAMILY:** Amaryllidaceae | **HEIGHT:** 50cm | **SPREAD:** 10cm | **EXPOSURE:** Full sun | **ASPECT:** South or west facing

The summer snowflake looks very like a snowdrop, with the same dainty white flowers marked with green at the tips of its petals, but at 50cm tall, the snowflake towers above its relation. It's also later flowering, with up to eight bell-shaped flowers on each stem, appearing between March and May. It's easy to grow, but needs moist soil. In the right conditions snowflakes will naturalise throughout borders and will also flourish around the edges of ponds. Plant the bulbs in autumn, around 10cm deep.

30 MARCH

Wild daffodil *Narcissus pseudonarcissus*

PLANT TYPE: Bulb | **FAMILY:** Amaryllidaceae | **HEIGHT:** 35cm | **SPREAD:** 10cm | **EXPOSURE:** Full sun, Partial shade | **ASPECT:** South, north, east or west facing

To tell a wild daffodil from a garden variety growing in parks and alongside roads, look for a smaller daffodil with two-tone colouring: pale yellow petals and a darker yellow trumpet. The long, thin leaves are a grey-green colour. Until the 19th century the wild species was common: most daffodils were either wild species or natural hybrids, but it is now far harder to spot the wild species, most often found in ancient woodland. The wild daffodil is also known as the Lent lily.

31 MARCH

Witch hazel *Hamamelis mollis* 'Goldcrest'

PLANT TYPE: Deciduous shrub | **FAMILY:** Hamamelidaceae | **HEIGHT:** 4.5m | **SPREAD:** 4.5m | **EXPOSURE:** Dappled shade, Partial shade, Full sun | **ASPECT:** South, north, east or west facing

'Goldcrest' has yellow flowers with a dark-red centre, which are larger than the blooms you'd find on the standard species. These appear before the leaves and are sweetly scented. Chinese witch hazels are easy-to-grow, reliable, winter-flowering shrubs, prized for their spidery flowers, which come in shades of yellow, orange and red. A witch hazel makes a striking focal point for a winter border and grows best in well-drained, slightly acidic soil. Plant it in autumn while the soil is still warm to give it time to establish.

APRIL

1 APRIL

Guelder rose *Viburnum opulus*

PLANT TYPE: Deciduous shrub | **FAMILY:** Viburnaceae | **HEIGHT:** 8m | **SPREAD:** 4m | **EXPOSURE:** Full sun, Partial shade | **ASPECT:** South, north, east or west facing

The Guelder rose is a significant plant in the Ukraine and features in many folk songs. It has three seasons of interest, with white lacecap flowers from late spring to early summer, autumn berries and bright red autumn colour. This native shrub is an indicator of ancient woodland, and can also be found alongside rivers, as it thrives in moist well-drained soil. It's also a valuable plant for wildlife – the bright-red berries provide food for birds, such as mistle thrushes and bullfinches, while the flowers are attractive to hoverflies, and it's also a caterpillar food plant.

2 APRIL

Striped squills *Puschkinia scilloides* var. *libanotica*

PLANT TYPE: Bulb | **FAMILY:** Asparagaceae | **HEIGHT:** 10cm | **SPREAD:** 5cm | **EXPOSURE:** Full sun, Partial shade | **ASPECT:** South, north, east or west facing

Also known as the Russian snowdrop, *Puschkinia scilloides* var. *libanotica* is more exotic-looking than a snowdrop, with up to ten star-shaped flowers on each flower spike, which have a blue stripe down each petal. Striped squills are suited to growing in rock gardens or beneath trees and shrubs – they do need some sun, so avoid planting below trees with a dense canopy. They are a good partner for snake's head fritillaries, anemone, iris or cyclamen. Plant in the autumn for flowers from March to April. The bulbs do best in a spot that is moist in spring but warm and dry in summer. If planting in the lawn, avoid mowing until June.

3 APRIL

Wild tulip *Tulipa sylvestris*

PLANT TYPE: Bulb | **FAMILY:** Liliaceae | **HEIGHT:** 25cm | **SPREAD:** 10cm | **EXPOSURE:** Full sun | **ASPECT:** South, east or west facing

This is a native tulip, which grows wild in the UK. Once you plant it in a garden it will be there for years, growing well in grass and under deciduous trees. It's a tough and reliable species tulip with scented yellow blooms on slim stems. *Tulipa sylvestris* naturalises easily and will multiply over the years to create attractive drifts. Plant late October to November once the weather is colder. It's a good pick for rockeries or containers and makes a good cut flower.

4 APRIL

Wood anemone *Anemone nemorosa* 'Blue Eyes'

PLANT TYPE: Perennial | **FAMILY:** Ranunculaceae | **HEIGHT:** 15cm | **SPREAD:** 15cm | **EXPOSURE:** Partial shade | **ASPECT:** South, north, east or west facing

The name 'Blue Eyes' refers to the centre of this eye-catching flower, which is a vivid blue. Like the species, 'Blue Eyes' is easy to grow and a good plant for filling gaps beneath trees and shrubs. This cultivar, which is a double-flowered variety, adds extra colour to shady areas, although it can also be grown in spring containers, with flowers between March and April. It's vigorous and low-growing, reaching only 15cm tall. Plant the rhizomes around 8cm deep from September to October.

---

5 APRIL

Grape hyacinth *Muscari armeniacum*

PLANT TYPE: Bulb | **FAMILY:** Asparagaceae | **HEIGHT:** 20cm | **SPREAD:** 10cm | **EXPOSURE:** Full sun, Partial shade | **ASPECT:** South, north, east or west facing

Muscari are easy bulbs as they're low maintenance and will grow almost anywhere, flowering between April and May. They spread easily, creating attractive groundcover for borders, rockeries and woodland areas, although this does mean clumps may need dividing every few years. Muscari also make pretty container plants partnered with miniature daffodils, pansies or tulips. This variety has simple blue flowers that will attract spring pollinators, including the hairy-footed flower bee, *Anthophora plumipes*.

6 APRIL

Lily of the valley *Convallaria majalis*

PLANT TYPE: Perennial | **FAMILY:** Asparagaceae | **HEIGHT:** 25m | **SPREAD:** 30cm | **EXPOSURE:** Partial shade, Full shade | **ASPECT:** North, east or west facing

In France, people give sprigs of lily of the valley to friends and family at the Fête du Muguet on 1 May, Labour Day, as a symbol of good luck. Lily of the valley is also a classic and popular addition to wedding bouquets. The white bell-shaped flowers are scented, held on arching stems, brightening up shady borders through April and May. Lily of the valley grows well in moist soil in partial shade, spreading to create attractive groundcover. There is also a pretty pink variety, *Convallaria majalis* var. *rosea*.

7 APRIL

Snake's head fritillary *Fritillaria meleagris*

PLANT TYPE: Bulb | **FAMILY:** Lilaceae | **HEIGHT:** 30cm | **SPREAD:** 5cm | **EXPOSURE:** Full sun, Partial shade | **ASPECT:** South, north, east or west facing

The unusual checked pattern of the flowers of *Fritillaria meleagris* is given as the reason for both its Latin and common names. The common name 'Snake's head fritillary' could be because the pattern looks like the head of snake; 'Fritillary' could come from the Latin '*fritillus*', meaning dice box (some Roman dice towers had a latticework pattern). It makes the flowers stand out in borders and containers. The decline in meadows in the UK is a threat to this plant, which used to be found in swathes in damp meadows. It grows best in soil that doesn't dry out during summer but also thrives in spring borders and containers.

8 APRIL

Candelabra primula *Primula allionii*

PLANT TYPE: Perennial | **FAMILY:** Primulaceae | **HEIGHT:** 10cm | **SPREAD:** 10cm | **EXPOSURE:** Full sun, Partial shade | **ASPECT:** South or east facing

These early-flowering primroses flower from February to April and usually come in various shades of pink and purple. These are Alpine primulas that grow in the European alps. They can be grown in rock gardens in rich, but well-drained soil, although some need to be kept in a cold greenhouse or alpine frame to protect the foliage, which grows in a rosette, from wet weather. There are lots of hybrids of this plant and these are often easier to look after than the species.

9 APRIL

Aubretia *Aubretia* 'Purple Cascade'

PLANT TYPE: Perennial | **FAMILY:** Brassicaceae | **HEIGHT:** 10cm | **SPREAD:** 40cm | **EXPOSURE:** Full sun | **ASPECT:** South or west facing

Aubretia is named after Claude Aubriet, a 17th-century French botanical artist, who was chief draughtsman for the Royal Botanic Garden in Paris and carried out commissions for King Louis XIV. It's a mat-forming perennial that produces a mass of small flowers in spring. This hardy perennial is ideal for growing in rockeries or over walls as it can tolerate dry conditions, but it also looks lovely at the front of a border or as underplanting in containers. 'Purple Cascade' has deep purple flowers from March to May and evergreen hairy leaves.

10 APRIL

Cowslip *Primula veris*

PLANT TYPE: Perennial | **FAMILY:** Primulaceae | **HEIGHT:** 25cm | **SPREAD:** 25cm | **EXPOSURE:** Full sun, Partial shade | **ASPECT:** South, north, east or west facing

Traditionally, cowslips are a plant of wildflower meadows and ancient woodlands, an unmistakable sign of spring. According to the charity Plantlife, it was once as common as the buttercup but suffered serious declines between 1930 and 1980 with the loss of grassland to modern farming. In the garden, these flowers will flourish in moist soil, such as near the edges of ponds or in a bog garden. The funnel-shaped, bright yellow flowers are held at the top of upright stems and attract bees and butterflies.

11 APRIL

Barrenwort *Epimedium lishihchenii*

PLANT TYPE: Perennial | **FAMILY:** Berberidaceae | **HEIGHT:** 45cm | **SPREAD:** 30cm | **EXPOSURE:** Partial shade | **ASPECT:** North, east or west facing

Having damp shade in your garden is a good excuse to plant epimediums, which will brighten gloomy borders and spread to suppress weeds. Epimediums are also known as Bishop's hat, because of the shape of their flowers, resembling a biretta (a cap worn by the clergy). *Epimedium lishihchenii* is a large species with yellow flowers and evergreen leaves that turn orange in the autumn. Epimediums grow best in acid soil. Mulch around plants with garden compost or leafmould. Cut back old leaves before new ones appear in the spring.

12 APRIL

Pheasant's eye *Narcissus poeticus*

PLANT TYPE: Bulb | **FAMILY:** Amaryllidaceae | **HEIGHT:** 35cm | **SPREAD:** 10cm | **EXPOSURE:** Full sun, Partial shade | **ASPECT:** South, east or west facing

Pheasant's eye daffodil is a late-flowering species that blooms through April and May. Its flowers have a strong scent and are easily recognisable with their pure white petals. The yellow cup is small and red-edged. *Narcissus poeticus* is a reliable old variety that naturalises easily in grass and looks impressive in large drifts, but can also be used in pots or borders alongside other native wildflowers such as fritillaries. The fragrance of these daffodils makes it the ideal variety to cut for an indoor display.

13 APRIL

Oxlip *Primula elatior*

PLANT TYPE: Perennial | **FAMILY:** Primulaceae | **HEIGHT:** 45cm | **SPREAD:** 45cm | **EXPOSURE:** Full sun, Partial shade | **ASPECT:** South, north, east or west facing

The oxlip is often found in ancient woodland, and predominantly in East Anglia where it grows in damp woods and meadow. It's the county flower of Suffolk and was traditionally used to cure coughs. According to The Woodland Trust it is now a rare species in the wild, although it is available to buy and will grow well in moist soil in woodland areas where it will naturalise. The bell-shaped yellow flowers appear from April to May – oxlips are often confused with cowslips, but these have darker-yellow flowers and are found growing throughout the UK.

14 APRIL

Mediterranean spurge *Euphorbia characias* subsp. *wulfenii*

PLANT TYPE: Shrub | **FAMILY:** Euphorbiceae | **HEIGHT:** 1.2m | **SPREAD:** 1.2m | **EXPOSURE:** Full sun | **ASPECT:** South, east or west facing

Large heads of vivid lime-green flowers give this plant a striking appearance between March and May, but this tall shrub also looks good year-round thanks to its evergreen foliage and architectural shape. It makes a refreshing change from traditional evergreen shrubs, well suited to both formal and contemporary gardens. Coming from the Mediterranean where it grows on dry ground, this euphorbia is able to tolerate drought and is happy in a sunny border or gravel garden. It's a good partner for plants such as verbena or red-hot pokers.

---

15 APRIL

Japanese quince *Chaenomeles japonica*

PLANT TYPE: Deciduous shrub | **FAMILY:** Rosaceae | **HEIGHT:** 1m | **SPREAD:** 2m | **EXPOSURE:** Full sun, Partial shade | **ASPECT:** South, north, east or west facing

Japanese quince is very easy to grow in most soils and can be trained as a wall shrub, which creates an attractive display of its cup-shaped flowers that are red to orange and loved by pollinators such as bees. Its leaves are round and glossy, with a hint of red before they mature and the plant's stems are thorny. The flowers are followed by autumn fruits, which can be made into quince jelly. This ornamental shrub is extremely hardy, down to -20°C. It makes a good low hedge and is sometimes used to create bonsai.

16 APRIL

Creeping blue blossom *Ceanothus thyrsiflorus* var. *repens*

PLANT TYPE: Evergreen shrub | **FAMILY:** Rhamnaceae | **HEIGHT:** 1m | **SPREAD:** 2.5m | **EXPOSURE:** Dappled shade, Full sun | **ASPECT:** South or west facing

Creeping blue blossom is from California, also known as Californian lilac, where the climate is more forgiving than ours. Although this variety is one of the hardiest ceanothus it's best planted in a sheltered spot, against a warm wall or in a mild garden. Throughout May and June, it's covered in a mass of pale-blue flowers that attract butterflies and bees. The plant forms a natural low mound with glossy evergreen leaves and makes an easy feature plant for beginners as it's low maintenance. It adds structure to borders and also grows well on a slope.

17 APRIL

Common name *Tulipa* 'Typhoon'

PLANT TYPE: Bulb | **FAMILY:** Liliaceae | **HEIGHT:** 50cm | **SPREAD:** 15cm | **EXPOSURE:** Full sun | **ASPECT:** South or west facing

Tulips are invaluable for spring colour, and with so many to choose from there's a colour and form to suit every border and container. There are traditional, single cup-shaped varieties, double-flowered types, parrot tulips with fringed petals. 'Typhoon' is a cup-shaped tulip with petals that form a point at the top. It's an elegant variety whose red and orange petals would suit a hot-themed border. Plant the bulbs in November, three times the depth of the bulb with the pointed end up.

18 APRIL

Armand clematis *Clematis armandii*

PLANT TYPE: Evergreen climber | **FAMILY:** Ranunculaceae | **HEIGHT:** 5m | **SPREAD:** 3m | **EXPOSURE:** Full sun | **ASPECT:** South or west facing

Clematis armandii was named after Father Armand David, a French missionary from 1826–1900 who travelled to China as a plant collector and naturalist. He was also the first Westerner to discover the Giant Panda. This evergreen clematis has almond-scented flowers in spring, but provides coverage for a wall or fence throughout the year. It's a good choice for planting near a doorway or entrance to appreciate its fragrance on warm spring days. Plant it in a border that doesn't get too wet as it grows best in free-draining soil. It only needs light pruning to keep it within bounds after flowering.

19 APRIL

Hawthorn *Crataegus monogyna*

PLANT TYPE: Deciduous tree | **FAMILY:** Rosaceae | **HEIGHT:** 8m | **SPREAD:** 8m | **EXPOSURE:** Full sun, Partial shade | **ASPECT:** South, north, east or west facing

The hawthorn was a Medieval symbol of fertility and has long associations with May Day festivities, when its leaves and flowers were used in May garlands. Despite its connections to May it begins flowering in April, with clusters of white blossom that are incredibly fragrant. It flowers best in full sun but will grow in most locations and its dense growth provides shelter for birds. It has benefits for a wide range of wildlife, providing food for caterpillars of moths, haws for birds and nectar for pollinating insects. As a hedge it will grow to around 1.5–3m in height, but it also makes a good tree for a small garden.

20 APRIL

Flagpole cherry, *Prunus × yedoensis*

PLANT TYPE: Tree | **FAMILY:** Rosaceae | **HEIGHT:** 12m | **SPREAD:** 8m | **EXPOSURE:** Full sun | **ASPECT:** South or west facing

Prunus × yedoensis is also known as the Tokyo cherry or Yoshino cherry due to its Japanese origins. It was grown for years in Yoshino and only introduced to Europe in 1902. This upright tree makes an elegant feature on a lawn where its spring blossom can be admired. It's not a one-season-only tree, as it also has attractive autumn colour, with the deep green leaves turning bright red and orange as the temperature drops. The flowers are white to light pink with a light scent, and are attractive to pollinators. The black fruits that follow are popular with birds.

21 APRIL

Chocolate vine *Akebia quinata*

PLANT TYPE: Climber | **FAMILY:** Lardizabalaceae | **HEIGHT:** 10m | **SPREAD:** 5m | **EXPOSURE:** Sun, Dappled shade | **ASPECT:** South, north, east or west facing

This is an interesting climber to cover a wall or fence, although your fence needs to be sturdy as the plant is vigorous and becomes heavy over the years. Its common name comes from its chocolate-coloured flowers, which have a spicy fragrance and appear in late spring. These are cup-shaped and often look more red-purple than brown. The foliage takes on more of a purple hue in winter and will fall unless it's particularly mild. There is also a white-flowered variety, 'Alba', which is pretty and looks attractive trained over an arch or pergola, where the flowers can hang down.

22 APRIL

Winter hazel *Corylopsis pauciflora*

PLANT TYPE: Shrub | **FAMILY:** Hamamelidaceae | **HEIGHT:** 1.5m | **SPREAD:** 2.5m | **EXPOSURE:** Partial shade | **ASPECT:** North, east or west facing

One of the common names for *Corylopsis pauciflora* is buttercup witch hazel, perhaps because it combines the pale yellow of buttercups with a delicate sweet scent. Clusters of bell-shaped flowers hang from bare branches in March and April. This slow-growing, multi-stemmed shrub thrives in acid soil in partial shade. It was introduced to Britain by the Scottish plant hunter Robert Fortune in 1860, who brought it back from Japan. The leaves have some golden autumn colour before they fall. Give this shrub space in the border as it grows up to 2.5m wide.

23 APRIL

Pieris *Pieris japonica* 'Purity'

PLANT TYPE: Evergreen shrub | **FAMILY:** Ericaceae | **HEIGHT:** 1.5m | **SPREAD:** 1.5m | **EXPOSURE:** Full sun, Partial shade | **ASPECT:** East or west facing

Pieris japonica 'Purity' brings structure to the border with its compact shape and evergreen foliage. Its highlight, though, is its lily-of-the-valley-style, bright white flowers in April and May. These are attractive to bees and stand out well against the deep-green leaves. Grow this shrub in partial shade in moist but well-drained acidic soil, preferably with some shelter from harsh winter winds. It goes well with other plants that thrive in acid soil, such as camellias and azaleas. If you don't have acid soil, it would be a good choice for a container, planted in ericaceous soil.

24 APRIL

Gordon's currant *Ribes × beatonii*

PLANT TYPE: Deciduous shrub | **FAMILY:** Grossulariaceae | **HEIGHT:** 2.5m | **SPREAD:** 2.5m | **EXPOSURE:** Full sun | **ASPECT:** South or west facing

Ribes × beatonii owes both its common names, Beaton's currant and Gordon's currant, to Donald Beaton, who raised it in 1837 and named it after his employer, William Gordon. Many currants have pink flowers but this ornamental currant has trusses of small red and yellow flowers in March and April. The flowers are nectar-rich and will attract pollinating insects such as bees and butterflies. Grow it in moist soil and prune annually to keep plants vigorous and bushy. This is a low-maintenance shrub that will look good with spring-flowering bulbs and perennials.

25 APRIL

Chilean iris *Libertia chilensis* Formosa Group

PLANT TYPE: Perennial | **FAMILY:** Iridaceae | **HEIGHT:** 90cm | **SPREAD:** 60cm | **EXPOSURE:** Full sun | **ASPECT:** South or west facing

The flowers of this architectural perennial last for weeks, well into June, making it a valuable part of a border in early summer. The white bowl-shaped flowers are held on upright stems and surrounded by evergreen sword-like leaves. After the flowers are over, attractive seed pods add interest and it will self-seed. It would make a good plant partner for grasses or hardy geraniums that begin flowering early. Divide congested plants in spring or autumn to rejuvenate them. This has the advantage of providing free plants to restock your borders or containers.

26 APRIL

Barberry *Berberis darwinii* 'Compacta'

PLANT TYPE: Evergreen shrub | **FAMILY:** Berberidaceae | **HEIGHT:** 90cm | **SPREAD:** 90cm | **EXPOSURE:** Full sun, Partial shade | **ASPECT:** South, north, east or west facing

Charles Darwin discovered *Berberis darwinii* in 1835 on his five-year voyage in the *Beagle*. It's native to Chile and Argentina and this compact form only grows to around 90cm tall. This makes it ideal for creating a low hedge, especially as it looks good year-round with its small, spiny evergreen leaves. It flowers in April and May with small orange to yellow flowers that are attractive to pollinators, and its blue berries provide food for birds. Plant it in full sun for the best display of flowers, in well-drained soil.

27 APRIL

Honesty *Lunaria annua*

PLANT TYPE: Hardy annual or biennial | **FAMILY:** Brassicaceae | **HEIGHT:** 90cm | **SPREAD:** 30cm | **EXPOSURE:** Full sun, Partial shade | **ASPECT:** South, east or west facing

Honesty has two highlights as a plant – its scented purple flowers and its round, pale, silvery seedheads which are popular with flower arrangers and remain attractive through winter, especially on cold frosty days. It can be grown as an annual or biennial and will self-seed freely if the seedheads are left in place over winter. It's very attractive to bees and other pollinators.

28 APRIL

Ornamental pear *Pyrus calleryana* 'Chanticleer'

PLANT TYPE: Deciduous tree | **FAMILY:** Roseaceae | **HEIGHT:** 12+m | **SPREAD:** 4–8m | **EXPOSURE:** Full sun | **ASPECT:** South, north, east or west facing

The ornamental pear, known as the callery pear, has a mass of single white flowers between March and May which are rich in nectar and pollen, valuable to bees, beneficial insects and other pollinators. Although it does produce fruit, the pears are inedible and shouldn't be eaten. 'Chanticleer' is a conical tree, reaching a height of around 12m when mature, after around 50 years. As well as attractive spring blossom, this ornamental pear has vibrant autumn colour, its leaves turning a vivid red before dropping.

29 APRIL

Victoria plum *Prunus domestica* 'Victoria'

PLANT TYPE: Deciduous tree | **FAMILY:** Rosaceae | **HEIGHT:** 4m | **SPREAD:** 4m | **EXPOSURE:** Full sun | **ASPECT:** South or west facing

One of the most popular plum trees due to its heavy cropping and reliable nature. The delicious fruit, which ripens in August to September, can be eaten raw or cooked and used in crumbles, chutneys and jams. It's self-fertile so doesn't need a pollination partner to produce fruit and has pretty blossom from April to May. It can reach about 4m in height, but can also be grown in a container on a dwarfing rootstock such as Pixy. For the best fruit, grow this tree in full sun, in free-draining soil.

---

30 APRIL

Rowan *Sorbus aucuparia*

PLANT TYPE: Deciduous tree | **FAMILY:** Rosaceae | **HEIGHT:** 15m | **SPREAD:** 7m | **EXPOSURE:** Full sun, Partial shade | **ASPECT:** South, north, east or west facing

Rowan trees can live for 200 years and were traditionally planted near homes to ward off misfortune, witches and spirits. It will grow in many types of soil but needs a large garden as it can reach 15m tall. The cream-white flowers are held in clusters during April and May and are followed by a mass of berries that attract a wide variety of birds including blackbirds, song thrush and mistle thrush. Also known as mountain ash, because it grows well at high altitude and is native to Northern Europe.

MAY

1 MAY

Bluebell *Hyacinthoides non-scripta*

PLANT TYPE: Bulb | **FAMILY:** Asparagaceae | **HEIGHT:** 40cm | **SPREAD:** 10cm | **EXPOSURE:** Partial shade | **ASPECT:** South, north, east or west facing

The UK has almost half the world's population of English bluebells. They are often a sign of ancient woodland and it is a criminal offence to uproot plants. Bluebells flower from April to May, naturalising in shady woodland and beneath trees. Their flowers are a violet blue and carry a sweet scent. Spanish bluebells have larger, paler flowers and are unscented. Bluebells should be planted 'in the green' as they will establish better after flowering when they still have leaves. If bluebells are growing in your lawn, avoid cutting that area until after the bluebell foliage has died down.

2 MAY

Ornamental onion *Allium stipitatum*

PLANT TYPE: Bulb | **FAMILY:** Alliaceae | **HEIGHT:** 50cm | **SPREAD:** 50cm | **EXPOSURE:** Full sun | **ASPECT:** South or west facing

Alliums add height and colour to the border with their globe-shaped flowerheads. *Allium stipitatum* has pink flowers from May to June that attract pollinators such as butterflies and moths. It mixes well with ornamental grasses, other varieties of allium, late tulips or early summer-flowering hardy geraniums. It makes a good cut flower, too, but leave some to develop seedheads, which will add architectural interest through the winter. Plant bulbs September to November in moist, well-drained soil in full sun.

3 MAY

Bearded iris *Iris* 'Sostenique'

PLANT TYPE: Bulb | **FAMILY:** Iridaceae | **HEIGHT:** 1m | **SPREAD:** 45cm | **EXPOSURE:** Full sun | **ASPECT:** South or west facing

Bearded irises are ideal for hot borders where they'll create a dramatic display in early summer. Bearded irises are named after the 'beards' or markings on the lower petals, which are also called 'falls'. The upright petals are called standards. *Iris* 'Sostenique' is a bearded iris that flowers in May and June in a sunny spot. It has orange beards, apricot pink standards and violet lower petals. Plant in autumn or spring, making sure the top of the rhizome is just above the soil surface so it gets the sun. If you are planting bare rhizomes, soak them briefly before planting.

4 MAY

Tulip 'Oviedo' *Tulipa* 'Oviedo'

PLANT TYPE: Tulip | **FAMILY:** Lilaceae | **HEIGHT:** 50cm | **SPREAD:** 20cm | **EXPOSURE:** Full sun | **ASPECT:** South or west facing

'Ovido' is a late-flowering, fringed tulip with pink and white ruffled petals. The base is white while the centre is pink – more of the pink centre is gradually revealed as the tulip ages and opens. Tulips look good in mixed displays with plants that thrive in the same conditions – sun and free-draining soil. Look for plants that complement the tulips' season of flowering and that can help disguise the tulips' foliage when they go over. Some popular planting companions, depending on a tulip's colour, include forget-me-nots, box, hebe and bluebells.

5 MAY

Wild hyacinth *Camassia leichtlinii* subsp. *leichtlinii*

PLANT TYPE: Bulb | **FAMILY:** Asparagaceae | **HEIGHT:** 1m | **SPREAD:** 20cm | **EXPOSURE:** Full sun, Partial shade | **ASPECT:** South, north, east or west facing

Camassias come from North America where the large bulbs were a food crop – the bulbs were used as a vegetable by Native Americans and either roasted or boiled. Now they are grown for their tall spikes of blue, star-shaped flowers in early summer, which at 1m tall are often grown in wildflower meadows or long grass. For a dramatic display it's grown in large drifts. Camassia is a long-lived bulb that's easy to grow and available in white, purple or blue varieties. It thrives in moist soil and over time it will multiply.

6 MAY

Granny's bonnet *Aquilegia flabellata*

PLANT TYPE: Perennial | **FAMILY:** Ranunculaceae | **HEIGHT:** 30cm | **SPREAD:** 15cm | **EXPOSURE:** Dappled shade | **ASPECT:** North, east or west facing

Aquilegias, also known as Columbine, have the common name Granny's bonnet due to the flowers resembling bonnets or hats. They are popular cottage-garden plants that self-seed around the garden readily and provide valuable nectar for bees. *Aquilegia flabellata* is a dwarf columbine, reaching up to 30cm, whereas other varieties can grow to triple this height. It has blue flowers from May to June and looks pretty at the front of a border or in a container display. Plant aquilegias in partial shade in moist soil.

7 MAY

Common name *Mathiasella bupleuroides* 'Green Dream'

PLANT TYPE: Perennial | **FAMILY:** Apiaceae | **HEIGHT:** 1m | **SPREAD:** 80cm | **EXPOSURE:** Full sun, Partial shade | **ASPECT:** South or west facing

Mathiasella bupleuroides 'Green Dream' was only discovered in 1954. This unusual perennial is native to Mexcico and is a compact cultivar of the species. The umbellifer flowerheads have clusters of green flowers, which are actually modified leaves. The actual flower is small and black, inside the green bracts. These small flowers hang down at the top of tall stems, which fade to pink over the season and last well into autumn with architectural appeal. It's a lovely plant for cutting to bring indoors. Plant it in moist, well-drained soil.

---

8 MAY

Oriental poppy *Papaver orientale*

PLANT TYPE: Perennial | **FAMILY:** Papaveraceae | **HEIGHT:** 90cm | **SPREAD:** 60cm | **EXPOSURE:** Full sun | **ASPECT:** South or west facing

The oriental poppy is a perennial that has large flowers, up to 15cm across, which have petals that range from flat to crimped or ruffled. Flowering in May and June, they're well suited to a spot at the front or middle of the border, alongside low-growing perennials that will hide the foliage when they go over. The species has bowl-shaped orange to red flowers with black patches at the base of each petal. Some popular cultivars include the 'Beauty of Livermere', which has large crimson scarlet flowers up to 20cm across.

---

9 MAY

Sicilian honey garlic *Nectaroscrodum siculum*

PLANT TYPE: Bulb | **FAMILY:** Amaryllidaceae | **HEIGHT:** 1.2m | **SPREAD:** 10cm | **EXPOSURE:** Dappled shade, Full sun | **ASPECT:** South, north, east or west facing

This striking plant is related to the allium and has the same architectural seedheads. The flowerheads are umbels on tall stems up to 1.2m tall. The small, creamy white flowers hang on short stalks and are flushed with a pretty red on the outer petals. This is an airy, graceful plant that can be dotted in among other plants without blocking the view or light. It's a beautiful choice for a gravel garden. *Nectaroscrodum siculum* can be grown from seed, but it will take a long time to get a flowering plant so it's better to plant bulbs in the autumn.

---

10 MAY

Herbaceous peony *Peonia lactiflora* 'Bowl of Beauty'

PLANT TYPE: Perennial | **FAMILY:** Paeoniaceae | **HEIGHT:** 1m | **SPREAD:** 80cm | **EXPOSURE:** Full sun, Partial shade | **ASPECT:** South, east or west facing

This peony is a true showstopper, the bowl of beauty in the name referring to the vibrant pink petals and ruffled-looking creamy centre. The middle of the flower is actually made up of lots of tiny petals. It's a bonus to the May border, adding a focal point before many of the summer stars get going in June. It is also scented. Herbaceous peonies come in many colours including white, pink, red and yellow, flowering in May and June. They die back to ground level in the winter.

11 MAY

Night-scented stock *Matthiola longipetala*

PLANT TYPE: Hardy annual | **FAMILY:** Brassicaeae | **HEIGHT:** 30cm | **SPREAD:** 30cm | **EXPOSURE:** Full sun | **ASPECT:** South or west facing

Some plants have a stronger scent at night so that they can attract moths as their main pollinators – night-scented stock is one of these, making it a good plant to grow to help conserve moths. Grow it in containers near a bench or doorway and it'll benefit you too, on a summer evening where it's warm enough to sit outside. The flowers are lilac and if sown in winter, spring and summer you'll have them in bloom from April through to September. Sow direct and generously for a good display.

12 MAY

Cranesbill *Geranium pratense* 'Mrs Kendall Clark'

PLANT TYPE: Perennial | **FAMILY:** Geraniaceae | **HEIGHT:** 90cm | **SPREAD:** 60cm | **EXPOSURE:** Full sun, Partial shade | **ASPECT:** South, east or west facing

Hardy geraniums are stalwarts of the summer border, combining reliability with beauty. They create useful groundcover between taller summer perennials and shrubs like roses, and help suppress weeds. 'Mrs Kendall Clark' was bred by nurseryman Walter Ingwersen in the 1930s and it's said to be named after a friend or customer who gave him the original plant. It's a tall, vigorous variety with blue-grey flowers that have pale white stripes. If cut back after flowering geraniums are likely to flower again.

13 MAY

Sea thrift *Armeria maritima*

PLANT TYPE: Perennial | **FAMILY:** Plumbaginaceae | **HEIGHT:** 50cm | **SPREAD:** 50cm | **EXPOSURE:** Full sun | **ASPECT:** South, north, east or west facing

In Orkney and the Outer Hebrides, thrift was not only an attractive coastal plant, but also used as a remedy for tuberculosis and a cure for hangovers. It also appeared as an emblem on the threepence coin in 1937, perhaps to remind people of the need to save money. Although found growing on cliffs and coastlines, it's also a pretty plant for gardens, where it will grow well on poor soil, if well-drained. Its pink globe-shaped flowers provide cheering colour between May and July and attract bees, butterflies and other pollinators.

14 MAY

Perennial wallflower *Erysimum* 'Bowles's Mauve'

PLANT TYPE: Perennial | **FAMILY:** Brassicaceae | **HEIGHT:** 75cm | **SPREAD:** 60cm | **EXPOSURE:** Full sun | **ASPECT:** South, east or west facing

Erysimum 'Bowles's Mauve' is prized for its long-flowering nature, it just keeps blooming, from late winter into autumn. Perhaps because of this vigorous flowering, it is a short-lived perennial, but it's easy to make new plants from cuttings, which can take over when your parent plant runs out of steam. The tall spires of purple flowers are a good match for spring-flowering bulbs such as tulips, as well as a later-flowering plants such as achillea or echinacea. Trim flower stalks as they fade to encourage more flowers.

15 MAY

Bear's breeches *Acanthus spinosus*

PLANT TYPE: Perennial | **FAMILY:** Acanthaceae | **HEIGHT:** 1.5m | **SPREAD:** 90cm | **EXPOSURE:** Full sun, Partial shade | **ASPECT:** South, north, east or west facing

Acanthus spinosus is dramatic in foliage and flower, with tall architectural flower spikes surrounded by spiny leathery leaves that can reach 90cm long. It has been used by the Greeks as a motif on roof ornaments and Corinthian columns. The flowers are white with purple hoods, held on tall stems, and these last for several weeks, although the plant should keep flowering from May to August. Place plants in their pots before planting to check the position is correct as it's difficult to move them once planted. They are best at the back of the border due to their height.

16 MAY

Solomon's seal *Polygonatum × hybridum*

PLANT TYPE: Perennial | **FAMILY:** Asparagaceae | **HEIGHT:** 90cm | **SPREAD:** 60cm | **EXPOSURE:** Full shade, Partial shade, Full sun | **ASPECT:** South, north, east or west facing

Solomon's seal has been a popular garden plant in Britain for centuries, perhaps because of the elegance of its white bell-shaped flowers. These are held on arching stems and brighten shady borders. The flowers last from May to June and have a light scent. This species is the most common, with attractive pleated, light-green leaves, but there are other varieties available with double flowers or variegated leaves. Grow it in moist, well-drained soil as it doesn't do well in dry conditions.

17 MAY

Sweet pea *Lathyrus odoratus*

PLANT TYPE: Hardy annual | **FAMILY:** Fabaceae | **HEIGHT:** 1.8m | **SPREAD:** 30cm | **EXPOSURE:** Full sun, Partial shade | **ASPECT:** South, east or west facing

The annual sweet peas climb using tendrils, covering structures such as wigwams and trellis with fragrant colourful blooms. The species is red and purple, growing up to around 2m, but there are hundreds of varieties to choose from, including some of the most popular, such as an old-fashioned type, 'Matucana', famous for its strongly scented, two-toned purple and magenta flowers. Sweet peas give a long season of flowers from May to August. Sow either in autumn or spring. They need to be tied in and watered regularly during summer.

---

18 MAY

Love-in-a-mist *Nigella damascena*

PLANT TYPE: Annual | **FAMILY:** Ranunculaceae | **HEIGHT:** 50cm | **SPREAD:** 20cm | **EXPOSURE:** Full sun | **ASPECT:** South or west facing

This hardy annual is easy to grow from seed and can be sown direct, where it should grow reliably with little intervention. If necessary, thin the plants to around 15cm apart to allow each plant room to develop well. The flowers are surrounded by a froth of lacey foliage, which is said to look like a cloud of mist, hence the common name. It has been a classic plant for cottage gardens since Elizabethan times. The flowers are followed by equally striking, large, alien-looking seedheads.

---

19 MAY

Sweet woodruff *Galium odoratum*

PLANT TYPE: Perennial | **FAMILY:** Rubiaceae | **HEIGHT:** 30cm | **SPREAD:** 1m | **EXPOSURE:** Dappled shade, Partial shade, Full sun | **ASPECT:** North, east or west facing

Galium odoratum is a woodland plant that is useful for groundcover between shrubs and trees, especially in shady areas. It's a fragrant plant, with star-shaped white flowers from mid-spring through to July. One plant can spread a metre or more. In Belgium and Germany sweet woodruff is used to flavour a drink called Maitrank. The dried leaves can also be used in syrups, teas and in potpourri. Plant sweet woodruff in moist, well-drained soil.

---

20 MAY

Sweet rocket *Hesperis matronalis*

PLANT TYPE: Biennial | **FAMILY:** Brassicaceae | **HEIGHT:** 90cm | **SPREAD:** 45cm | **EXPOSURE:** Full sun, Partial shade | **ASPECT:** South, north, east or west facing

Hesperis matronalis is known as Sweet rocket because of its scent, which is at its strongest in the evening and as sweet as a violet. This biennial flower is best suited to informal borders, where it can be sown in drifts to attract beneficial insects such as bees, butterflies and moths. The flowers are lilac but it can also come in shades from deep purple through to white. Some nurseries sell mixed packets. Sow the seed direct in late spring and leave some to self-seed to get plants for free.

21 MAY

Salvia 'Jezebel' *Salvia* 'Jezebel'

PLANT TYPE: Perennial | **FAMILY:** Lamiaceae | **HEIGHT:** 90cm | **SPREAD:** 30cm | **EXPOSURE:** Full sun | **ASPECT:** South or west facing

For bright summer colour, this showstopping salvia is hard to beat. It's prolific, flowering from May right through to November with vibrant red flowers. It's a popular plant with bees and makes an excellent cut flower. Grow it in a dry spot in full sun, with well-drained soil. Deadhead the flowers regularly to keep plants flowering into autumn. Don't prune after flowering as leaving the flowerheads on will help protect plants from frost. It can be cut back when new shoots emerge in the spring.

---

22 MAY

Bird cherry *Prunus padus*

PLANT TYPE: Deciduous tree | **FAMILY:** Rosaceae | **HEIGHT:** 15m | **SPREAD:** 10m | **EXPOSURE:** Full sun | **ASPECT:** South, north, east or west facing

Bird cherry flowers for a short time in spring, but the blooms provide an early source of nectar. They're almond scented and according to The Wildlife Trusts, honey made by bees visiting this tree is particularly delicious. The flowers are followed by bitter cherries that birds such as blackbirds and thrushes will eat. Bird cherry is a native tree, found commonly in northern and eastern parts of England, and is most suitable for a large garden, although there are compact varieties available.

23 MAY

Sea holly *Eryngium maritimum*

PLANT TYPE: Perennial | **FAMILY:** Apiaceae | **HEIGHT:** 50cm | **SPREAD:** 30cm | **EXPOSURE:** Full sun | **ASPECT:** South, east or west facing

This is the native sea holly, found growing in coastal areas where it thrives in free-draining soil, in a sunny spot. In the wild it can grow in sand and shingle, so it's also ideal for gravel gardens, where it will mingle well with ornamental grasses and perennials such as salvias or verbascums. Its leaves and flowers are blue and last over a long period from May to August. The spikey look of its flowerheads make it a popular choice for flower arrangements.

---

24 MAY

Twinspur *Diascia personata* 'Hopleys'

PLANT TYPE: Perennial | **FAMILY:** Scrophulariaceae | **HEIGHT:** 1m | **SPREAD:** 50cm | **EXPOSURE:** Full sun | **ASPECT:** South or west facing

This perennial is worth growing for its long-flowering nature, as it adds colours to borders from May through to September, sometimes even into late autumn. The bright pink flowers are held on foxglove-style spires that suits cottage-garden-style planting and it combines well with silver-leaved foliage plants such as stachys as well as other cottage-garden favourites such as nepeta or vibrant euphorbias, which flower around the same time. It's borderline hardy and benefits from a thick layer of mulch in winter to protect plants from winter wet and cold.

---

25 MAY

White laceflower *Orlaya grandiflora*

PLANT TYPE: Annual | **FAMILY:** Apiaceae | **HEIGHT:** 60cm | **SPREAD:** 30cm | **EXPOSURE:** Full sun | **ASPECT:** South facing

White laceflower comes from the Mediterranean where it grows wild in olive groves, fields and vineyards. Its umbellifer flower heads float above fern-like foliage, to create an elegant display. Grow them in among grasses or roses as a contrast to their white delicate flowers. This annual is especially attractive to hoverflies, providing nectar over a long period from May to August depending on when the seed is sown. Sow in autumn for earlier flowers the next year and overwinter seedlings in a coldframe.

26 MAY

Lilac *Syringa vulgaris*

PLANT TYPE: Deciduous shrub | **FAMILY:** Oleaceae | **HEIGHT:** 7m | **SPREAD:** 7m | **EXPOSURE:** Full sun | **ASPECT:** South, north, east or west facing

Lilac was a favourite of the Edwardians before going out of fashion, but there's much to love about this plant. Its famous for its scent, which is often used in perfume and soap. The flowers are light blue to purple with a wonderful fragrance. The species has given rise to many varieties in shades of pink, purple and white that usually flower between May and July. For something compact, 'Red Pixie' is easy to grow, with a mass of fragrant pink flowers, and as it grows to around 1.8m it is suitable for a container.

27 MAY

Million bells *Calibrachoa*

PLANT TYPE: Annual | **FAMILY:** Solanaceae | **HEIGHT:** 50cm | **SPREAD:** 50cm | **EXPOSURE:** Full sun | **ASPECT:** South or west facing

Calibrachoa is often referred to as million bells, although this is a trademark name. Native to South America, this annual is popular for its colourful, petunia-style flowers that last all through the summer. Most varieties tend to spread and trail rather than grow up, which makes them an excellent option for hanging baskets and containers where they can spill over the edge. They can't be grown from seed as they're hybrids, so they need to be grown from plug plants in the spring. Place them in full sun for the best display of flowers.

28 MAY

Snapdragon *Antirrhinum majus* 'Royal Bride'

PLANT TYPE: Short-lived perennial | **FAMILY:** Plantaginaceae | **HEIGHT:** 90cm | **SPREAD:** 45cm | **EXPOSURE:** Full sun | **ASPECT:** South or west facing

The stories behind the name snapdragon often involve a reference to a mouth. In some accounts it's because the flower snaps open when you pinch it, like the mouth of a dragon, but the botanical name *Antirrhinum* also translates roughly as 'like a snout or nose'. 'Royal Bride' flowers until September with pure white flowers on tall spikes up to 90cm tall. Snapdragons are traditionally used in cottage gardens and look pretty in a mixed border. This variety also has a scent, which is unusual for snapdragons and its flowers make good cut flowers. It's especially attractive to bumblebees.

---

29 MAY

Dusky cranesbill *Geranium phaeum*

PLANT TYPE: Perennial | **FAMILY:** Geraniaceae | **HEIGHT:** 90cm | **SPREAD:** 50cm | **EXPOSURE:** Dappled shade, Partial shade, Full shade, Full sun| **ASPECT:** South, north, east or west facing

This is an early-flowering cranesbill, with purple to almost black flowers that have white centres in May to June. Flourishing in sun or shade, this is a versatile plant for a border, and in shade will go well with foxgloves, ferns and epimediums, as well as thriving beneath deciduous trees. The foliage sometimes has distinctive purple blotches. This early-flowering species has many attractive cultivars including 'Album', a white variety, and 'Samobor', which will grow well in dry or damp shade and has small purple flowers from late spring to summer.

30 MAY

Common laburnum *Laburnum anagyroides*

PLANT TYPE: Deciduous trees | **FAMILY:** Fabaceae | **HEIGHT:** 8m | **SPREAD:** 8m | **EXPOSURE:** Full sun | **ASPECT:** South, north, east or west facing

Laburnum anagyroides is also called the golden rain tree, because its bright yellow flowers hang down on long stems. These racemes (stems) can reach 45cm long, with blooms all along their length. In big gardens they are commonly trained over an arch, so that the pea-like flowers cascade down through the structure. It's fast-growing, putting on around 40cm a year, but can be kept to a reasonable size with pruning. Although this tree can grow to 8m, there are varieties suitable for a small garden, such as 'Yellow Rocket', which has an upright growth habit and reaches 2m tall.

31 MAY

Wingpod purslane *Portulaca umbraticola*

PLANT TYPE: Succulent | **FAMILY:** Portulacaceae | **HEIGHT:** 15cm | **SPREAD:** 60cm | **EXPOSURE:** Full sun | **ASPECT:** South or west facing

The fleshy leaves and stems of this succulent are eaten as a vegetable in some countries. It's native to South and North America. It's also grown for its vibrant flowers that come in shades of orange, reds and pinks. It's easy to grow from seed and ideal for planting out into containers for a summer display where they make excellent drought-tolerant plants for hot summers right through to September. Cut back untidy growth in late summer to encourage new flower shoots.

JUNE

1 JUNE

Middendorf weigela *Weigela middendorffiana*

PLANT TYPE: Deciduous shrub | **FAMILY:** Caprifoliaceae | **HEIGHT:** 1m | **SPREAD:** 1m | **EXPOSURE:** Full sun, Partial shade | **ASPECT:** South, north, east or west facing

Weigelas are ornamental shrubs that flower in early summer. They're easy to grow, low maintenance and compact, which makes them a useful shrub for a mixed border. Native to Japan and northern China, weigela are hardy in the UK and grow well in moist, well-drained soil. The species *Weigela middendorffiana* has pale yellow, scented flowers from May to July, with darker markings in yellow-orange to yellow-red inside. Its leaves are a fresh, deep green. You'll often see bees inside the funnel-shaped blooms.

2 JUNE

Chinese wisteria *Wisteria sinensis*

PLANT TYPE: Climber | **FAMILY:** Fabaceae | **HEIGHT:** 9m | **SPREAD:** 4m | **EXPOSURE:** Full sun, Partial shade | **ASPECT:** South or west facing

Wisteria is often grown on the front of houses or over arches where its fragrant purple flowers transform walls in early summer. One of the most popular gardening queries must be, 'How do I get my wisteria to flower?' as they can be slow to establish and they may take a few years to begin flowering after planting. They also need to be pruned twice a year to keep them flowering well. The maintenance is worth it for the flowering display. The species has purple flowers but there are also white varieties as well as several in numerous shades of violet.

3 JUNE

Dwarf red larkspur *Delphinium nudicaule* 'Redcap'

PLANT TYPE: Perennial | **FAMILY:** Ranunculaceae | **HEIGHT:** 30cm | **SPREAD:** 30cm | **EXPOSURE:** Dappled shade, Full sun | **ASPECT:** South or west facing

Delphinium comes from the Greek word *delphinos*, meaning dolphin, thought to be perhaps due to the shape of its flowers. Its common name, larkspur, has been in use since Tudor times. This is a dwarf variety, only reaching 30cm tall. *Delphinium nudicaule* is a short-lived perennial that is native to California and flowers June to August with small bright orange flowers. The flowers are nectar-rich and attract bees. It's primarily a summer plant as the foliage dies down in the autumn but is perfect for a spot at the front of the border or in a container.

4 JUNE

Sea kale *Crambe maritima*

PLANT TYPE: Perennial | **FAMILY:** Brassicaceae | **HEIGHT:** 75cm | **SPREAD:** 60cm | **EXPOSURE:** Full sun | **ASPECT:** South, east or west facing

Sea kale is not only an attractive ornamental plant with large grey-green leaves, it's also a tasty vegetable. Both the stems and young leaves can be eaten raw or steamed. It can also be forced, in the same way you would force rhubarb to produce young, tender stems. It's often found growing on shingle beaches, especially around Suffolk and Essex, producing clusters of honey-scented white flowers in early summer to around July, that attract pollinating insects. This tough plant grows best in full sun.

5 JUNE

Common foxglove *Digitalis purpurea*

PLANT TYPE: Biennial | **FAMILY:** Plantaginaceae | **HEIGHT:** 1.5m | **SPREAD:** 45cm | **EXPOSURE:** Full sun, Partial shade | **ASPECT:** South, east or west facing

Foxgloves are poisonous – ingesting any part of the plant can have toxic effects due to them containing cardiac glycosides. This has also been used to help people, with the cardiac glycosides used to make a drug for heart problems. *Digitalis purpurea* is native to Western Europe, where it flourishes in woodland. Tall spires of purple flowers with a spotted throat appear in early summer from June to July. Bees will crawl inside the tubular flowers to get to the pollen. Although it's biennial, if the flowerheads are left to drop their seeds, you should get new plants for free.

6 JUNE

Tree peony *Paeonia × suffruticosa*

PLANT TYPE: Shrub | **FAMILY:** Paeoniaceae | **HEIGHT:** 2m | **SPREAD:** 2m | **EXPOSURE:** Full sun, Partial shade | **ASPECT:** South or west facing

The name 'tree peonies' is misleading, as these plants are actually small shrubs, but unlike herbaceous peonies, which die back, the woody framework can be seen year round. The goblet-shaped flowers are in bloom through May and June. The species is native to China – an upright shrub with pink flowers, but its cultivars come in many shades from white through to purple, including a pink and white bi-coloured variety, 'Dao Jin'. The flowerheads can be up to 30cm across. Deadhead after flowering.

7 JUNE

Masterwort *Astrantia major* 'Hadspen Blood'

PLANT TYPE: Perennial | **FAMILY:** Apiaceae | **HEIGHT:** 60cm | **SPREAD:** 45cm | **EXPOSURE:** Full sun, Partial shade | **ASPECT:** South, north, east or west facing

Astrantia is also known as Hattie's pincushion, and although there's no reliable source for who Hattie is, the 'pincushion' part is reliably connected to the flowerheads. The flowers of 'Hadspen Blood' are a deep red, given added stand-out value by the dark plant stems. It's a dramatic addition for areas beneath trees and shrubs, as well as moist borders. After flowering, rejuvenate plants by cutting them back to the ground – this will encourage new foliage and flowers to develop. Astrantia can grow in drier ground, but will benefit from a layer of mulch around them in spring.

---

8 JUNE

Green foxglove *Digitalis viridiflora*

PLANT TYPE: Perennial | **FAMILY:** Plantaginaceae | **HEIGHT:** 90cm | **SPREAD:** 30cm | **EXPOSURE:** Full sun, Partial shade | **ASPECT:** South, north, east or west facing

This is a perennial species of foxglove, unusual because of its green-yellow flowers, hence the name green foxglove. It's native to countries such as Greece, Bulgaria, Turkey and Croatia and is harder to get hold of in the UK than other foxgloves. It flowers May to June and grows best in partial shade, to replicate the growing conditions that most foxgloves thrive in – moist, well-drained soil. If you don't want seedlings, cut back the flower spike once it has faded. This can encourage new side shoots and flowers.

---

9 JUNE

Tree lupin *Lupinus arboreus*

PLANT TYPE: Evergreen shrub | **FAMILY:** Fabaceae | **HEIGHT:** 2m | **SPREAD:** 2m | **EXPOSURE:** Full sun, Partial shade | **ASPECT:** South or west facing

This tree lupin comes from California and likes a spot in full sun where possible, for the best display of its pea-like yellow flowers. Sometimes it may even have blue flowers, but this is rare. Its blooms are nectar, and pollen-rich, attracting bees, butterflies and moths over its long flowering period from May through to August. As well as bright cheery flowers, it has evergreen foliage and is fast growing, reaching 2m tall. This makes it a good choice for the middle of a border.

---

10 JUNE

Salvia *Salvia nemorosa* 'Caradonna'

PLANT TYPE: Perennial | **FAMILY:** Lamiaceae | **HEIGHT:** 50cm | **SPREAD:** 30cm | **EXPOSURE:** Full sun, Partial shade | **ASPECT:** South, east or west facing

'Caradonna' is a clump-forming salvia with tall spikes of purple flowers that are loved by bees and other pollinators. It flowers from June through to October, a beautiful plant for ornamental borders. Plant alongside other flower shapes for contrast, such as hardy geraniums, alliums or daisy-style flowers, or add in a colour contrast with silver-foliage plants like stachys. Plant in a sunny spot, in well-drained soil. Keep deadheading to encourage flowering into mid autumn.

11 JUNE

Yarrow *Achillea millefolium*

PLANT TYPE: Perennial | **FAMILY:** Asteraceae | **HEIGHT:** 60cm | **SPREAD:** 60cm | **EXPOSURE:** Full sun | **ASPECT:** South or west facing

This perennial wildflower looks beautiful in wildlife gardens or informal borders where its flat-topped flowerheads create a sea of white. The umbel-type flowerheads look striking mixed in with contrasting shapes such as the spikes of salvias or the pincushion-style flowers of scabious. Hoverflies, bees and butterflies are attracted to its blooms, which appear from June to August. There are many vibrant yarrow varieties in shades including yellow, bright pink and terracotta orange.

12 JUNE

Sage 'Amistad' *Salvia* 'Amistad'

PLANT TYPE: Perennial | **FAMILY:** Lamiaceae | **HEIGHT:** 1.5m | **SPREAD:** 75cm | **EXPOSURE:** Full sun | **ASPECT:** South or west facing

This striking salvia can be in flower for months, from June through to November, especially if deadheaded regularly. It's easy to grow, with rich purple, velvety flowers and almost black stems for extra dramatic impact. Salvias are versatile ornamental plants that can look good in a range of planting schemes, from cottage garden to tropical and contemporary. This variety is very popular with bees. Grow it in a sunny spot in well-drained soil and deadhead regularly to encourage flowering well into autumn. Don't cut back the flowers until spring as the old blooms will help protect the plant from frost.

13 JUNE

Sneezeweed *Helenium* 'Moerheim Beauty'

PLANT TYPE: Perennial | **FAMILY:** Asteraceae | **HEIGHT:** 90cm | **SPREAD:** 60cm | **EXPOSURE:** Full sun | **ASPECT:** South, east or west facing

The intriguing common name sneezeweed is down to the fact that the dried leaves and heads of helenium were used to make snuff, and this caused sneezing. This would apparently rid the body of evil spirits. Heleniums flower from June through to August and are popular in prairie-style gardens, planted alongside ornamental grasses. Their vibrant red and orange flowers go well with other hot-coloured plants such as rudbeckias, crocosmia and dahlias but can equally be planted alongside cooler-toned plants such as nepeta or salvias. 'Moerheim Beauty' has crimson-red flowers and thrives in sunny borders.

14 JUNE

Small scabious *Scabiosa columbaria*

PLANT TYPE: Perennial | **FAMILY:** Caprifoliaceae | **HEIGHT:** 75cm | **SPREAD:** 50cm | **EXPOSURE:** Full sun | **ASPECT:** South, north, east or west facing

Scabious is such a pretty flower but the meaning behind its name is less attractive – it's thought to come from the Latin *scabere*, which means to scratch, as it was used to treat scabies in Medieval times. The pincushion flowers of this species are a beautiful blue to violet shade and, like all scabious, is a bee magnet, blooming from June through to October. It thrives in poor soil so will grow well in a gravel garden, but also looks good in a container and makes an excellent cut flower. Deadhead regularly to keep the flowers coming.

15 JUNE

Pot marigold *Calendula*

PLANT TYPE: Hardy annual | **FAMILY:** Asteraceae | **HEIGHT:** 50cm | **SPREAD:** 30cm | **EXPOSURE:** Full sun, Partial shade | **ASPECT:** South, east or west facing

Calendula is an incredibly useful plant – it's a companion plant for vegetables as it helps attract pollinating insects, the petals are edible and it is used in skincare products. Calendula cream can be used for mild sunburn, itchy skin and minor skin rashes. The flowers are a vivid orange, bringing colour to borders and containers over the summer from June right through until October. This cottage-garden favourite is easy to grow. Sow the seeds direct in spring and keep seedlings well watered.

16 JUNE

Field poppy *Papaver rhoeas*

PLANT TYPE: Hardy annual | **FAMILY:** Papaveraceae | **HEIGHT:** 90cm | **SPREAD:** 30cm | **EXPOSURE:** Full sun | **ASPECT:** South or west facing

The field poppy is well known as the symbol of Remembrance Day, with its bright red flowers that appear between June and August. It's often seen growing wild in fields or on the side of roads but it is also a valuable addition to gardens, growing easily from seed and attracting bumblebees. It thrives in sun, adding colour to wildflower meadows or in a mixed border among summer-flowering perennials. Sow the seed direct onto well-prepared soil, which has had all stones and weeds removed.

17 JUNE

Rose deutzia *Deutzia × rosea*

PLANT TYPE: Deciduous shrub | **FAMILY:** Hydrangeaceae | **HEIGHT:** 1m | **SPREAD:** 1m | **EXPOSURE:** Full sun, Partial shade | **ASPECT:** South, east or west facing

Deutzia are hardy, easy shrubs that aren't fussy about soil types. *Deutzia × rosea* also has the benefit of being compact, perfect for gardens that need to add structure, with star-shaped white flowers that have a pink flush on the outside of the petals. They have a short flowering season in early summer, around May to June, so it's a good idea to plant it alongside summer-flowering perennials and underplant with bulbs to extend the season of interest. Prune after flowering, to keep the shrub in good shape and encourage plenty of flowers the following year.

---

18 JUNE

Hydrangea *Hydrangea arborescens* 'Annabelle'

PLANT TYPE: Deciduous shrub | **FAMILY:** Hydrangeaceae | **HEIGHT:** 2.5m | **SPREAD:** 2.5m | **EXPOSURE:** Full sun, Partial shade | **ASPECT:** South, north, east or west facing

Large snowball-shaped flowers make this an ideal feature plant, perfect as a focal point in a border. The flowerheads can be up to 30cm across and last until around September. They are also often used as dried flowerhead in cut flower arrangements. In autumn the flowers fade to a lime colour. Leave the flowerheads on over winter to protect the new buds from frost. Young plants may need support as the flowerheads are heavy, but mature plants will develop stronger branches.

---

19 JUNE

Mock orange *Philadelphus* 'White Rock'

PLANT TYPE: Deciduous shrub | **FAMILY:** Hydrangeaceae | **HEIGHT:** 3m | **SPREAD:** 2.5m | **EXPOSURE:** Full sun, dappled shade | **ASPECT:** South or west facing

The scented flowers of philadelphus are said to smell and look like orange tree blossom, which is why it has the name 'mock orange'. This shrub has spectacular pure white flowers from April to June on upright plants that need little maintenance. Its fresh look is ideal for an early summer border, alongside roses or plants such as peonies, which are also a highlight at this time of year. Philadelphus grows best in moist, well-drained soil in full sun. Keep this shrub in shape by cutting back to a strong pair of buds every year after flowering.

---

20 JUNE

Heliotrope *Heliotropium arborescens*

PLANT TYPE: Shrub (grown as an annual) | **FAMILY:** Boraginaceae | **HEIGHT:** 50cm | **SPREAD:** 50cm | **EXPOSURE:** Full sun | **ASPECT:** South or west facing

In South America heliotrope is grown as a perennial shrub, but it's half hardy in the UK and so is mostly used as annual summer bedding. Its compact shape and fragrant flowers make it ideal for container displays in late spring to early summer. The flowers are especially attractive to butterflies. *Heliotropium arborescens* has violet or lavender-blue flowers and in a pot will grow to around 45cm. In the ground it may reach around a metre tall. It has many cultivars that are used for summer bedding, including 'Chatsworth', which is strongly scented with deep-purple flowers.

21 JUNE

Common elder *Sambucus nigra*

PLANT TYPE: Shrub or tree | **FAMILY:** Viburnaceae | **HEIGHT:** 6m | **SPREAD:** 6m | **EXPOSURE:** Full sun, Partial shade | **ASPECT:** South, north, east or west facing

The flowers of common elder can be used to make cordial or gin. They have a musky scent, and the airy, large, flat flowerheads are made up of lots of tiny white flowers. Elder is often seen growing in hedgerows, flowering from late May to July, but can also be used in a mixed native hedge. Elder were once used to keep away evil spirits, and it's said they were planted outside doorways to keep the devil away. Cut back hard in spring to encourage large leaves and a shrubby growing habit or leave it to grow if you want it to grow into a tree.

22 JUNE

Jasmine *Jasminum officinale*

PLANT TYPE: Deciduous climber | **FAMILY:** Oleaceae | **HEIGHT:** 12m | **SPREAD:** 3m | **EXPOSURE:** Full sun, Partial shade | **ASPECT:** South or west facing

Jasmine will cover fences and walls with clusters of sweetly scented flowers from June to August. The flowers are star-shaped and held in clusters along the stems. Although this pretty climber will tolerate shade, it grows best in full sun, on a south- or west-facing wall. It's a vigorous climber, clinging with its twining stems. This makes it a useful shrub if you have a large boundary to cover or an ugly shed or building to disguise. This species is also used to make jasmine tea.

23 JUNE

Sweet alyssum *Lobularia maritima* 'Snowdrift'

PLANT TYPE: Annual | **FAMILY:** Brassicaceae | **HEIGHT:** 10cm | **SPREAD:** 15cm | **EXPOSURE:** Full sun | **ASPECT:** South or east facing

The mass of white flowers that appears on low mounds of evergreen foliage does look a bit like the snowdrift in this plant's name. Sweet alyssum is perfect for planting alongside paths, because of this low-growing habit, although it is also lovely in a container combined with other summer annuals such as lobelias. As it's attractive to pollinating and beneficial insects it's also useful as an edging plant around veg beds to encourage insects that will pollinate flowers and help control pests. It flowers for months from June until October.

24 JUNE

Viper's bugloss *Echium vulgare*

PLANT TYPE: Biennial | **FAMILY:** Boraginaceae | **HEIGHT:** 90cm | **SPREAD:** 30cm | **EXPOSURE:** Full sun | **ASPECT:** South, north, east or west facing

This British native flower has collected some interesting common names, including adderwort, blue devil and cat's tail. It's most commonly known as viper's bugloss, said to be because its spotted stems resemble a viper. The spikes of blue and pink flowers are particularly attractive to bees because they continue to produce nectar throughout the day. It's visited by buff-tailed and red-tailed bumblebees as well as honeybees and red mason bees, flowering from May sometimes into August. The species is biennial and can be grown from seed, but one of its cultivars, 'Blue Bedder', is grown as an annual.

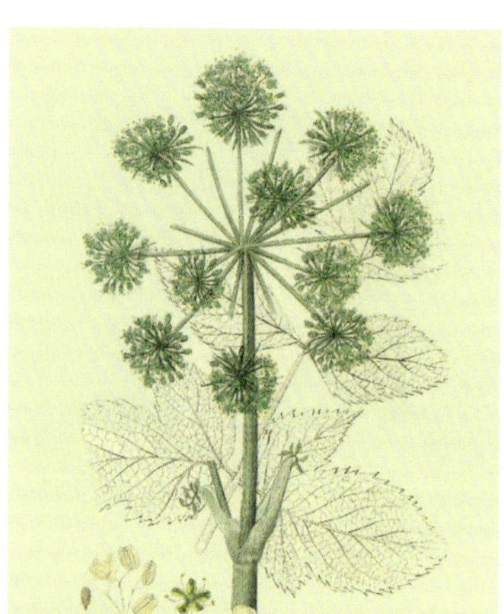

25 JUNE

Angelica *Angelica archangelica*

PLANT TYPE: Biennial | **FAMILY:** Apiaceae | **HEIGHT:** 2m | **SPREAD:** 1.2m | **EXPOSURE:** Full sun, Partial shade | **ASPECT:** North or east facing

Angelica is grown for its architectural flowerheads – large umbels up to 25cm across of small green flowers that resemble cow parsley. These plants look particularly striking planted at the back of a border and have pink stems that stand out in contrast to its flowers. The flowers are followed by pretty seedpods. It mixes well with plants like rodgersia or ferns. As well as its decorative attributes, angelica is also used to make candied stems for cake decorations and the leaves can be used as a natural sweetener. Grow it in moist soil, in partial shade.

26 JUNE

Globeflower *Trollius europaeus*

PLANT TYPE: Perennial | **FAMILY:** Ranunculaceae | **HEIGHT:** 80cm | **SPREAD:** 40cm | **EXPOSURE:** Full sun, Partial shade | **ASPECT:** South, east or west facing

Globeflower is a member of the buttercup family and looks like a giant buttercup with its rounded lemon-yellow flowerheads. It grows naturally in damp meadows, thriving in heavy, moist soil that doesn't dry out. It's a good choice for the edge of a pond or bog garden, although the ground should be moist rather than waterlogged and it needs plenty of sun. This makes it a useful plant for problem areas such as clay soil. Its round flowers provide welcome colour from May to July.

---

27 JUNE

Tickseed *Coreopsis verticillata* 'Zagreb'

PLANT TYPE: Perennial | **FAMILY:** Asteraceae | **HEIGHT:** 45cm | **SPREAD:** 45cm | **EXPOSURE:** Full sun, Partial shade | **ASPECT:** South, north, east or west facing

Coreopsis is a wonderful plant for wildlife gardens, attracting bees and butterflies. It's also reliable and long flowering with a mass of bright yellow daisy flowers in June and July that will boost borders. Perfect for filling gaps in the border or adding to summer containers, 'Zagreb' is an award-winning variety with large flowers, around 2.5cm across, held on wiry stems above filigree foliage. It goes well with grasses such as *Stipa tenuissima* or other perennials in flower at this time, such as salvias or nepeta.

28 JUNE

Water forget-me-not *Myosotis scorpioides*

PLANT TYPE: Marginal aquatic perennial | **FAMILY:** Boraginaceae | **HEIGHT:** 30cm | **SPREAD:** 30cm | **EXPOSURE:** Full sun, Partial shade | **ASPECT:** South, east or west facing

Garden forget-me-nots tend to be biennial, but water forget-me-not is perennial and will increase over the years. It's a pretty wildlife plant to have at the edge of a pond, benefitting a range of creatures including bees and hoverflies. It provides shelter for tadpoles and female newts use the leaves for their fertilised eggs. Its ornamental value comes from its tiny blue flowers that bloom from May to July. It can be planted in a basket or direct in the soil in shallow water, but no deeper than 10cm. Cut back plants after flowering.

29 JUNE

Greater sea kale *Crambe cordifolia*

PLANT TYPE: Perennial | **FAMILY:** Brassicaceae | **HEIGHT:** 2m | **SPREAD:** 1.5m | **EXPOSURE:** Full sun, Partial shade | **ASPECT:** South, east or west facing

Greater sea kale towers at the back of borders, a mass of tiny white flowers held on tall branching stems. The stems grow out of a mound of large leaves and it needs a lot of space as its spread can be up to 1.5m. One of Gertrude Jekyll's favourite plants, *Crambe cordifolia* is an asset to borders as its profusion of tiny scented flowers provide a contrast to other flowering perennials. Grow this plant in moist, fertile and well-drained soil – it needs a deep border because it has a long tap root.

---

30 JUNE

Pearl bush *Exochorda* x *macrantha* 'The Bride'

PLANT TYPE: Deciduous shrub | **FAMILY:** Rosaceae | **HEIGHT:** 2m | **SPREAD:** 3m | **EXPOSURE:** Full sun, Partial shade | **ASPECT:** South, north, east or west facing

This deciduous shrub is said to have the name pearl bush because when the flower buds are opening it looks like a string of pearls is hanging around the branches. *Exochorda* x *macrantha* 'The Bride' is a beautiful feature plant in May and June, with arching stems that are covered with pure white flowers. Although it looks good on its own, it's easy to extend its season of interest by growing a clematis such as *Clematis alpina* through it, which starts flowering before the shrub.

JULY

1 JULY

Passion flower *Passiflora caerulea*

PLANT TYPE: Climber | **FAMILY:** Passifloraceae | **HEIGHT:** 12m | **SPREAD:** 4m | **EXPOSURE:** Full sun, Partial shade | **ASPECT:** South or west facing

Passion flower is an evergreen climber with exotic, bowl-shaped flowers from July until September. Bees love these colourful flowerheads with a purple, blue and white centre, made up of lots of slender filaments. This vigorous climber is self-clinging but needs a trellis or wires on a fence to climb up. The flowers are followed by egg-shaped, orange fruits, but these are better left as food for the birds as they aren't very tasty. The more popular passion fruit is produced by the species *Passiflora edulis*. Plant in full sun if possible, although it will grow in partial shade.

2 JULY

Star of Persia *Allium cristophii*

PLANT TYPE: Bulb | **FAMILY:** Alliaceae | **HEIGHT:** 60cm | **SPREAD:** 20cm | **EXPOSURE:** Full sun | **ASPECT:** South or west facing

Star of Persia seems an apt name for this bulb, which has flowerheads made up of around 80 tiny, star-shaped, purple flowers. These create a striking globe that is a highlight of early summer borders, spectacular in amongst euphorbias or ornamental grasses. *Allium cristophii* also makes a good cut flower, but if left to stand in the border the seedheads will provide architectural interest well into autumn. Plant bulbs in September to October in a spot that gets plenty of sun, for flowers from June to July. Bulbs should be planted at a depth that is twice as deep as the bulb.

3 JULY

Short-stalked catmint *Nepeta subsessilis*

PLANT TYPE: Perennial | **FAMILY:** Lamiaceae | **HEIGHT:** 90cm | **SPREAD:** 30cm | **EXPOSURE:** Full sun | **ASPECT:** South, east or west facing

As its name suggests, this nepeta is more compact than many catmints and has an upright habit. Nepeta is renowned for being a magnet for bees and butterflies and this species is no different. It has large pale purple flowers that appear from July to September. Each flower on the spike has a small lower lip that is often spotted. This species is a good pick for the front of a border or containers. Grow it in moist, well-drained soil and prune after flowering to encourage bushy growth. If clumps get congested they can be divided in spring to create new plants.

4 JULY

Vial's primrose *Primula vialii*

PLANT TYPE: Perennial | **FAMILY:** Primulaceae | **HEIGHT:** 50cm | **SPREAD:** 30cm | **EXPOSURE:** Partial shade | **ASPECT:** North, east or west facing

This dramatic plant will give a lift to areas in the garden where there's some shade. Its bi-colour flowers are lilac-pink at the base with a bright red top, like a red hot poker. It was named *Primula littonia* by the Scottish plant hunter George Forrest, after his friend, although it was later discovered that it had already been named *P. vialii* by another plant hunter, a French missionary. These summer-flowering primulas grow in damp alpine meadows in Southern China. They are ideal for a pond edge or bog garden providing a bright contrast in June and July to plants like ferns and hostas.

5 JULY

Spider flower *Cleome houtteana* 'Helen Campbell'

PLANT TYPE: Annual | **FAMILY:** Cleomaceae | **HEIGHT:** 1.2m | **SPREAD:** 50cm | **EXPOSURE:** Full sun | **ASPECT:** South or west facing

The common name for cleome, spider flower, is down to its very long stamens – the combination of these and its long seed pods in the autumn give these flowers a spidery look. They have large scented flowers. 'Helen Campbell' is a white variety that mixes well in an ornamental border with other annuals such as cosmos and amaranthus. It flowers July to September and looks beautiful at the back of a sunny border where the tall plants add height and elegance. Sow the seed in early spring and plant outside once the danger of frost is over.

6 JULY

Pineapple lily *Eucomis autumnalis*

PLANT TYPE: Bulb | **FAMILY:** Asparagaceae | **HEIGHT:** 40cm | **SPREAD:** 20cm | **EXPOSURE:** Full sun | **ASPECT:** South, east or west facing

Plant this summer-flowering bulb in spring, from March to May, for eye-catching flowers from July to August. *Eucomis* is a South African flower that likes a sunny spot in the border. Its common name, Pineapple lily, comes from its appearance – the spikes of tiny greenish-white flowers are topped by a spiky tuft of leaves. These bulbs will also look good in a container. There are many other types of pineapple lily to choose from, including *Eucomis comosa* 'Sparkling Burgundy', which has purple stems and flowers that are pink with a hint of purple.

7 JULY

Penstemon 'Geoff Hamilton' *Penstemon 'Geoff Hamilton'*

PLANT TYPE: Perennial | **FAMILY:** Plantaginaceae | **HEIGHT:** 90cm | **SPREAD:** 60cm | **EXPOSURE:** Full shade, Partial shade | **ASPECT:** South, east or west facing

Penstemons are native to North America and became known in Europe as botanists began to explore the continent in the 1800s and 1900s. They became very popular in Britain with the Victorians, who used them as bedding plants. Their appeal lies in their foxglove-shaped flowers and long flowering period, sometimes blooming from July through to October or November. 'Geoff Hamilton' has ruby red flowers and was named after the popular *Gardeners' World* TV presenter. Mulch around plants in the autumn to protect the roots from cold weather.

---

8 JULY

Common hogweed *Heracleum sphondylium*

PLANT TYPE: Biennial | **FAMILY:** Apiaceae | **HEIGHT:** 2m | **SPREAD:** 50cm | **EXPOSURE:** Full sun, Partial shade | **ASPECT:** South, north, east or west facing

Common hogweed can be seen on the edges of roads, hedgerows or allotments. It's often classed as a weed but it's also a wildflower with large, flat-topped flowerheads that attract a huge range of insects, including bees. The flowerheads can reach 20cm across and its leaves can reach 60cm long. It's sometimes taken for cow parsley, but it can be identified by its leaves, which are not as finely divided. The dry, hollow stems are sometimes used by solitary bees when flowering is over. The sap is toxic, although not as toxic as that of giant hogweed. It's still advisable to wear gloves when handling the plant, though.

---

9 JULY

Argentinian vervain *Verbena bonariensis*

PLANT TYPE: Perennial | **FAMILY:** Verbenaceae | **HEIGHT:** 1.5m | **SPREAD:** 45cm | **EXPOSURE:** Full sun | **ASPECT:** South or west facing

Verbena bonariensis is an airy plant that adds height to a border with its flat-topped purple flowers. Its slim, stiff stems don't add bulk to a planting display, which makes this plant a versatile addition to an ornamental border – despite its height, it can be dotted through the middle as well as the back of a display without blocking other plants. Its flowering period is long, from June to September, with the nectar-rich blooms attracting bees and butterflies. Plant it in a sheltered spot, with well-drained soil. Plants will self-seed and the dead stalks can be left for interest over winter.

---

10 JULY

Hollyhock *Alcea rosea* 'Sunshine'

PLANT TYPE: Biennial/Short-lived perennial | **FAMILY:** Malvaceae | **HEIGHT:** 1.8m | **SPREAD:** 50cm | **EXPOSURE:** Full sun | **ASPECT:** South or west facing

Although native to China, Hollyhocks have been a favourite of British cottage gardens for centuries, towering plants with large open flowers all the way up their tall stems. *Alcea rosea* 'Sunshine' has large, single yellow flowers from June to September and is part of the Spotlight series of hollyhocks. This group of hollyhocks are perennial rather than biennial, although it's sometimes best to sow hollyhocks fresh each year, as mature plants can suffer from hollyhock rust. Hollyhocks also self-seed freely, so new plants may appear with little effort involved. They are easy to grow from seed in the spring, thriving in most types of soil.

11 JULY

Purple coneflower *Echinacea purpurea*

PLANT TYPE: Perennial | **FAMILY:** Asteraceae | **HEIGHT:** 1.5m | **SPREAD:** 60cm | **EXPOSURE:** Full sun, Partial shade | **ASPECT:** South, east or west facing

Echinacea is well known for its use in herbal medicine – herbalists recommend it for boosting the immune system and to help fight off colds. It's also popular in the garden, used in cottage-garden and prairie-style displays, mixing well with plants such as rudbeckias and ornamental grasses. The pink flowers have a central, orange-brown cone, hence the common name 'coneflower', and the stems are sturdy enough not to need staking. This is a resilient plant that's valuable in the border as it can withstand bad weather, it's easy to grow, the flowers are long-lived and it's attractive to pollinators.

12 JULY

Fennel *Foeniculum vulgare*

PLANT TYPE: Perennial | **FAMILY:** Apiaceae | **HEIGHT:** 1.8m | **SPREAD:** 60cm | **EXPOSURE:** Full sun | **ASPECT:** South or west facing

Fennel is ornamental and edible – its plate-like yellow flowers provide an attractive display during July and August, as does the feathery green foliage, but it's also an herb. The leaves, bulb and seeds can all be eaten. The leaves have a light aniseed flavour, while the seeds are often used in curries. The seeds are also popular with some birds and the flowers attract hoverflies and other insects. Sow the seed in spring. After flowering, the plant will die down and regrow the following spring. *Foeniculum vulgare* is related to the vegetable fennel, but this plant is perennial and will grow back each year.

13 JULY

Gentian sage *Salvia patens*

PLANT TYPE: Perennial | **FAMILY:** Lamiaceae | **HEIGHT:** 60m | **SPREAD:** 45cm | **EXPOSURE:** Full sun | **ASPECT:** East facing

Salvia patens is a tender perennial with vibrant blue flowers all through the summer. It's a stalwart of summer herbaceous borders because it's easy to grow from seed and has a long flowering period from June or July to October. In mild regions it may be possible to leave plants in the ground with some protection but in cold areas it's best to lift the plants and store them as you would with dahlias, until the spring. There are attractive cultivars available, including 'Cambridge Blue', which has pale blue flowers.

14 JULY

California poppy *Eschscholzia californica*

PLANT TYPE: Annual | **FAMILY:** Papaveraceae | **HEIGHT:** 30cm | **SPREAD:** 15cm | **EXPOSURE:** Full sun | **ASPECT:** South or west facing

These cheery annuals are the official state flower of California, native to Mexico and the south of the USA. They are invaluable for brightening up areas of poor soil that other plants struggle to thrive in, with bright flowers in shades including orange, yellow and red. Although California poppies are technically short-lived perennials, they are grown as annuals in the UK. Easily sown from seed, these poppies can be sown direct outdoors in a spot that gets full sun. They are a good choice for coastal gardens, wildflower meadows and dry borders.

---

15 JULY

Bishop's flower *Ammi majus*

PLANT TYPE: Annual | **FAMILY:** Apiaceae | **HEIGHT:** 1m | **SPREAD:** 50cm | **EXPOSURE:** Dappled shade, Partial shade, Full sun | **ASPECT:** South, east or west facing

Ammi majus has delicate, graceful flowers, similar to cow parsley. Its elegant flowers hover in a white haze between other plants, making an attractive contrast to more robust colourful perennials. Bishop's flower is popular for bouquets and cut flowers, as they can last up to ten days. Its flowers appear from June to August and the seedheads provide food for birds such as goldfinches. Sow indoors in February to March or direct in September to October. Sowing in the autumn will produce larger and taller plants than those sown in the spring. Plants may need staking.

16 JULY

Cornflower *Centaurea montana*

PLANT TYPE: Perennial | **FAMILY:** Asteraceae | **HEIGHT:** 45cm | **SPREAD:** 60cm | **EXPOSURE:** Dappled shade, Full sun | **ASPECT:** South, north, east or west facing

Centaureas are said to be named after the centaur Chiron in Greek mythology, who discovered that these plants had healing properties. Cornflowers are still used in herbal medicine today. *Centaurea montana* is also a tough but pretty border plant, which is easy to grow and hardy. Its blue to violet spidery flowers with a thistle-like centre appear above a spreading clump of foliage from May sometimes into September. Remove any faded flowers to encourage another flush of flowering in the autumn.

17 JULY

Bear's breeches *Acanthus mollis*

PLANT TYPE: Perennial | **FAMILY:** Acanthaceae | **HEIGHT:** 1.5m | **SPREAD:** 90cm | **EXPOSURE:** Full sun, Partial shade | **ASPECT:** South, east or west facing

Acanthus mollis makes a bold architectural plant for the back of a border, with tall flower spikes on stems over a metre tall. The flowers appear during July and August and stand out because of the hooded purple bracts covering the white flowers. These attract bees and are also useful for flower arrangements, both fresh and dried. Acanthus grows well in shady areas but is also drought-tolerant and will thrive in most soils. Its drawback is that it will form a clump that is hard to move once established – any bits of the plant that are left behind will form new plants.

---

18 JULY

Garden nasturtium *Tropaeolum majus*

PLANT TYPE: Annual | **FAMILY:** Tropaeolaceae | **HEIGHT:** 2.5m | **SPREAD:** 2.5m | **EXPOSURE:** Full sun | **ASPECT:** South or west facing

This annual climber has yellow, orange and red flowers all through summer into September, and can grow to 3m long, scrambling up supports such as wigwams or trellis. Nasturtiums are often used as a companion plant, to attract small and large white butterflies away from brassica crops. The flowers can also be used in salads. For a variety with a trailing habit, try 'Orange Troika', which has dark orange-red flowers and is more compact, growing to only 30cm. This would make a good nasturtium for a container, softening the edges of the pot with its trailing stems.

---

19 JULY

Opium poppy *Papaver somniferum*

PLANT TYPE: Annual | **FAMILY:** Papaveraceae | **HEIGHT:** 1.2m | **SPREAD:** 30cm | **EXPOSURE:** Full sun, Partial shade | **ASPECT:** South, east or west facing

Opium poppies have single or double flowers in shades such as purple, white, red or mauve. Although an individual flower doesn't last long, there's a profusion of blooms from June to August. The large seedheads are also attractive, and often used dried in flower arrangements. Opium poppies are native to Turkey. As suggested from the name, opium can be made using this plant – from a fluid that comes from the unripe seed pod. Sow the seed March to May, direct outdoors. Leave seedheads after flowering and the plants will self-seed around your garden.

---

20 JULY

Cosmos 'Purity' *Cosmos bipinnatus* 'Purity'

PLANT TYPE: Annual | **FAMILY:** Asteraceae | **HEIGHT:** 2.5–4m | **SPREAD:** 60cm | **EXPOSURE:** Full sun, Partial shade | **ASPECT:** South, east or west facing

Cosmos is a great-value annual to grow, because it flowers prolifically from July through to September. Its height makes it a good planting partner for dahlias and chrysanthemums, ideal for filling gaps in the border. Cosmos is also an elegant flower for containers or for vases indoors, where it lasts well as a cut flower. 'Purity' is a white, single-flowered variety but there are many to choose from, including double-flowered and those in shades including pink, red, orange and yellow. Deadhead regularly to keep the flowers coming. It's easy to grow, sow seed from March to April – it can be sown indoors and planted out in May or June, or sown direct outdoors.

21 JULY

Sunflower *Helianthus annuus*

PLANT TYPE: Annual | **FAMILY:** Asteraceae | **HEIGHT:** 3m | **SPREAD:** 45cm | **EXPOSURE:** Full sun | **ASPECT:** South or west facing

Sunflowers are easy to grow, with cheery flowerheads that perk up the back of the border for months on end. Sunflowers are in bloom from July to October, after which it's worth leaving the seedheads over winter for the birds. There's a huge range of varieties to choose from including the giant 'Titan', which c an grow to 3m, as well as compact types like 'Micro Sun', which only reaches 60cm. Both of these are a sunny yellow. Sow sunflowers indoors from March to May in small pots. Once they have germinated, pot on into larger pots before planting out into the garden.

22 JULY

Meadowsweet *Filipendula ulmaria*

PLANT TYPE: Perennial | **FAMILY:** Rosaceae | **HEIGHT:** 90cm | **SPREAD:** 60cm | **EXPOSURE:** Full sun, Partial shade | **ASPECT:** South, north, east or west facing

Meadowsweet is a native plant that thrives in damp grass and woodland and is often found growing next to rivers or in meadows. It's an important food plant for moths, including the Mottled Beauty and Emperor Moths. Clusters of tiny white to cream flowers appear from June to September, on tall stems around 1m tall, and these attract a wide variety of insects. Meadowsweet has one of the same ingredients as aspirin and was often used steeped in tea as a painkiller before aspirin and other painkillers were widely available.

23 JULY

Yellow sage *Lantana camara*

PLANT TYPE: Evergreen shrub | **FAMILY:** *Verbenaceae* | **HEIGHT:** 1.5m | **SPREAD:** 1.5m | **EXPOSURE:** Full sun | **ASPECT:** South facing

Each flowerhead of *Lantana camara* is made up of lots of flowers, which on the species are usually yellow, maturing to red. There are also white, pink and red types, including a pretty, more compact variety 'Lemon Cream'. After pollination the flowers change colour – they might go from yellow to red, for example. This is a sign to pollinating insects to find nectar in the yellow flowers. This shrub is native to tropical areas of America, which means it thrives best in moist soil in sunny borders. It's half hardy so it's best grown in a container that can be moved indoors during cold periods. It's also a popular conservatory plant.

---

24 JULY

Hops *Humulus lupulus*

PLANT TYPE: Perennial climber | **FAMILY:** Cannabaceae | **HEIGHT:** 5m | **SPREAD:** 1.8m | **EXPOSURE:** Dappled shade, Full sun | **ASPECT:** South, north, east or west facing

The hops that are used to make beer develop from the flowers of this deciduous climber. It's in flower from July through to September – the female flowers look like catkins and are green-yellow, whereas the male flowers are in a loose spray. It's the female flowers that are used to develop hops for beer. Hop flowers are also used in dried flower arrangements. This climber is fast growing and also makes an attractive covering for an arch or pergola. Plant it in moist, well-drained soil in full sun. Cut back foliage every spring to encourage new growth.

---

25 JULY

Lavender *Lavandula angustifolia* 'Hidcote'

PLANT TYPE: Evergreen shrub | **FAMILY:** Lamiaceae | **HEIGHT:** 60cm | **SPREAD:** 75cm | **EXPOSURE:** Full sun | **ASPECT:** South, east or west facing

English lavender is hardier than the French varieties but it does need to be planted in well-drained soil to thrive. This compact shrub has fragrant evergreen foliage and spikes of purple flowers, which are attractive to bees, butterflies and hoverflies. There are also white and pink varieties. Planted along a path or in a container, lavender adds summer fragrance to the garden. There are many beautiful cultivars to choose from, but for a small garden 'Hidcote' is a popular variety, taking its name from the garden of plantsman Laurence Johnston. It's a lovely lavender for edging paths.

26 JULY

Melancholy thistle *Cirsium heterophyllum*

PLANT TYPE: Perennial | **FAMILY:** Asteraceae | **HEIGHT:** 1.1m | **SPREAD:** 80cm | **EXPOSURE:** Full sun, Partial shade | **ASPECT:** South, north, east or west facing

This native plant gained the name Melancholy thistle because it was used as a cure for 'melancholia', or depression. It's referenced as being used in wine in Nicholas Culpeper's *Complete Herbal*. Its natural habitat is upland meadows, but its spiky purple-pink flowers are useful for brightening borders from July to August. It makes a good plant partner for hardy geraniums and alliums with its contrasting flower shape and tall stems. The flowers are loved by bees and the seedheads, left over winter, will provide food for birds. Rejuvenate plants by dividing them every few years.

27 JULY

Orange ball tree *Buddleia globosa*

PLANT TYPE: Deciduous shrub | **FAMILY:** Scrophulariaceae | **HEIGHT:** 5m | **SPREAD:** 5m | **EXPOSURE:** Full sun, Partial shade | **ASPECT:** South, east or west facing

This buddleia offers something different from the familiar purple form. *Buddleia globosa* has small, orange, ball-shaped flowers held at the end of long stems from June until August. These are scented and attractive to bees, butterflies and other pollinators. It's native to Chile, Peru and Argentina and benefits from a warm spot in the shelter of wall. Prune after flowering to encourage new shoots, then each year remove around a quarter of the old shoots down to the base to prevent the plant becoming leggy.

28 JULY

Myrtle *Myrtus communis*

PLANT TYPE: Evergreen shrub | **FAMILY:** Myrtaceae | **HEIGHT:** 3m | **SPREAD:** 3m | **EXPOSURE:** Full sun | **ASPECT:** South or west facing

Myrtle has an illustrious history, its sprigs used by the ancient Greeks to crown winners at the Olympic Games and since Queen Victoria's era, it has been a traditional addition to royal brides' bouquets. This beautiful shrub has been grown in Britain since the 16th century and makes an attractive hedge or shrub with glossy evergreen leaves and scented, fluffy-looking white flowers from July to August. The flowers are followed by dark purple berries. Native to the Mediterranean, myrtle benefits from a warm, sheltered spot in full sun, especially in colder regions of the UK. It's also an attractive container plant.

---

29 JULY

Corncockle *Agrostemma githago*

PLANT TYPE: Annual | **FAMILY:** Caryophyllaceae | **HEIGHT:** 60cm | **SPREAD:** 60cm | **EXPOSURE:** Full sun | **ASPECT:** South or west facing

Corncockle used to be common in cornfields, but it's now unlikely to be seen in the wild. The seeds of corncockle are poisonous, so it's no bad thing that it's only grown as an ornamental these days. As a garden flower, it has pretty pink flowers from June to August, often sold as a wildflower meadow plant along with poppies and cornflowers. For early June flowers, sow it the previous August, direct where you want it to grow. Otherwise, sow in March to April.

30 JULY

Corn marigold *Glebionis segetum*

PLANT TYPE: Annual | **FAMILY:** Asteraceae | **HEIGHT:** 50cm | **SPREAD:** 30cm | **EXPOSURE:** Full sun | **ASPECT:** South or west facing

Corn marigold is another annual that used to be seen growing in fields. It was so common that it was regarded as a weed in arable crop fields, which is why it has the name corn marigold – in common with corn poppy and corncockle. It is now often sold in wildflower meadow mixes, popular for its golden, daisy-like flowers that last from June through to September. Its flowers are beneficial to pollinators. Sow the seed direct in the spring, where you want it to flower. It used to be known as *Chrysanthemum segetum* but is now called *Glebionis segetum* and is sold under both names.

31 JULY

Ox-eye daisy *Leucanthemum vulgare*

PLANT TYPE: Perennial | **FAMILY:** Asteraceae | **HEIGHT:** 40cm | **SPREAD:** 30cm | **EXPOSURE:** Full sun, Partial shade | **ASPECT:** South, north, east or west facing

Ox-eye daisy is a native plant, the largest of the UK's daisies with big flowers on tall, slim stems. It can be found growing in meadows or on the edges of roads, but it also makes a useful ornamental plant, flowering from May through to September. It's valuable to pollinators such as hoverflies and bees because the centre of the plant is made up of lots of nectar-rich flowers. The game of picking petals, 'he loves me, he loves me not' is traditionally played using daisies, especially ox-eye daisies. It originated in France and was called '*effeuiller la marguerite*'.

AUGUST

1 AUGUST

Ivy-leaved cyclamen *Cyclamen hederifolium*

PLANT TYPE: Bulb | **FAMILY:** Primulaceae | **HEIGHT:** 10cm | **SPREAD:** 15cm | **EXPOSURE:** Dappled shade, Partial shade, Full shade | **ASPECT:** South, north, east or west facing

Also known as ivy-leafed cyclamen, the delicate-looking pink flowers appear in the autumn, ideal for lighting up areas under deciduous trees. It will self-seed to make bigger drifts of flowers each year. The dark green, silver-lined leaves of *Cyclamen hederifolium* are also a highlight, with the marbled heart-shaped foliage creating attractive groundcover after the flowers have finished. Plant the bulbs in September to October, in humus-rich soil. Mulch each year with well-rotted leaf mould – this will protect the tubers from heat in summer and the cold in the winter.

2 AUGUST

Dahlia 'Oakwood Goldcrest'

PLANT TYPE: Perennial | **FAMILY:** Asteraceae | **HEIGHT:** 1.1m | **SPREAD:** 30cm | **EXPOSURE:** Full sun | **ASPECT:** South or west facing

Dahlias provide late-summer colour in herbaceous borders and containers, and with such a range of shapes and colours to choose from, there should be a variety to suit everyone. Most will flower from July through to at least September. 'Oakwood Goldcrest' is a tall semi-cactus dahlia with pale yellow, pointed flowers. It's a good idea to stake plants early in their growing season to stop them toppling over. Deadhead regularly to encourage more flowers to develop. In cold regions it may be necessary to lift and store the tubers over winter. Pot them up and keep them in a cool, dry and frost-free place until March.

3 AUGUST

Dahlia 'Moor Place'

PLANT TYPE: Perennial | **FAMILY:** Asteraceae | **HEIGHT:** 1.2m | **SPREAD:** 40cm | **EXPOSURE:** Full sun | **ASPECT:** South or west facing

Dahlias are native to Mexico, where they were declared the National Flower in 1963. These flowers were originally grown by the Aztecs for their tubers, which they used as a food crop. They were introduced into Europe by Spanish colonists in the 18th century. 'Moor Place' is a pompon variety with incredible round, maroon-purple flowers. Pompon dahlias have petals that curve inwards to create perfect spheres and their flowers are smaller than ball dahlias, at about 5cm across. Dahlias can be planted out in the garden in mid to late May, once there is no danger of frost.

4 AUGUST

Dahlia 'Christopher Taylor'

PLANT TYPE: Perennial | **FAMILY:** Asteraceae | **HEIGHT:** 1.2m | **SPREAD:** 60cm | **EXPOSURE:** Full sun | **ASPECT:** South or west facing

'Christopher Taylor' is a crimson-flowered, waterlily dahlia. This type of dahlia has shallow, flat and double flowers that are shaped like a saucer with curved or flat petals. The flowerheads of waterlily dahlias are around 15cm across, nothing compared to the giant flowerheads of the plate dahlia, which can reach 30cm across. Dahlias are a versatile border plant, mixing well with grasses, cosmos, cannas and *Verbena bonariensis*, as well as many other summer-flowering perennials. They suit many styles of garden, from exotic hot borders to a traditional cottage garden.

---

5 AUGUST

Waterlily *Nymphaea odorata* subsp. *tuberosa*

PLANT TYPE: Aquatic perennial | **FAMILY:** Nymphaeaceae | **HEIGHT:** 10cm | **SPREAD:** 1.2m | **EXPOSURE:** Full sun | **ASPECT:** South or west facing

This hardy waterlily is a good choice for a lake or large pond – its spread is around 1.2m once established, with large, white, cup-shaped flowers. These have a yellow centre and appear from June through to September. Waterlilies are beneficial to wildlife, providing shade for fish, shelter for frogs and nectar for insects. At first, plant waterlilies so that they that have around 15–20cm water above the basket, placing them on a pile of bricks or similar support. Lower the basket gradually over the growing season. Remove any faded foliage in the autumn.

6 AUGUST

Iceplant *Hylotelephium spectabile*

PLANT TYPE: Perennial | **FAMILY:** Crassulaceae | **HEIGHT:** 45cm | **SPREAD:** 45cm | **EXPOSURE:** Full sun | **ASPECT:** South or east facing

Most commonly known by its former name, *Sedum,* the iceplant is grown for its flat flowerheads that add late colour to the border. The flowers are held above rosettes of succulent foliage from August to October. The masses of pink flowers are especially attractive to late-flying butterflies and bees in autumn, when fewer plants are in flower. This plant is easy to fit into a planting scheme, complementing a range of perennials such as geraniums, stachys, rudbeckias and asters. For a good display, grow sedums in full sun, in well-drained soil. Divide clumps every few years to improve flowering.

7 AUGUST

True London pride *Saxifraga umbrosa*

PLANT TYPE: Perennial | **FAMILY:** Saxifragaceae | **HEIGHT:** 30cm | **SPREAD:** 50cm | **EXPOSURE:** Full shade, Partial shade | **ASPECT:** South, north, east or west facing

London pride is said to have got this name because it appeared and flourished on bomb sites during World War II. However, *Saxifraga* × *urbium*, which is the result of a cross between *S. umbrosa* and *S. spathularis*, also has the common name London pride, so it may be that *S.* × *urbium* is the true hero of Noel Coward's song 'London Pride', which features the flower. Either way, *Saxifraga umbrosa* is a plant to be proud of in any garden, with pale pink, star-shaped flowers appearing above fleshy foliage. It's a good plant for a shady spot at the front of the border or a rockery.

---

8 AUGUST

Short-leaved stonecrop *Sedum brevifolium*

PLANT TYPE: Cactus or succulent | **FAMILY:** Crassulaceae | **HEIGHT:** 10cm | **SPREAD:** 50cm | **EXPOSURE:** Full sun | **ASPECT:** South, east or west facing

The short-leaved stonecrop is a lovely plant for rock gardens or creating a container display. Its tiny succulent leaves can be grey, green, blue-green, often with red tints, and in winter it stands out with its colourful evergreen foliage. The star-shaped, small white flowers appear at the end of the branches in summer, around June to August. This is a good succulent to grow in a container with a mulch of pebbles to show off the plant and would make a good companion for other hardy succulents if grown outdoors. It also grows well in the cracks and crevices of walls.

---

9 AUGUST

Hydrangea 'Le Vasterival'
Hydrangea paniculata 'Le Vasterival'

PLANT TYPE: Shrub | **FAMILY:** Hydrangeaceae | **HEIGHT:** 1.5m | **SPREAD:** 1m | **EXPOSURE:** Full sun, Partial shade | **ASPECT:** South or west facing

'Le Vasterival' is sometimes listed under the name 'Great Star'. It has panicles of white flowers, with graceful, fluted petals between July and September, which fade to deep pink in late autumn. This lovely shrub is a great choice for a feature plant, with fragrant flowers, an attractive structure in the border as well as being wildlife friendly. Leave the faded flowerheads until the spring, then prune. This will help protect the new buds from frost damage. It may be necessary to remove a few of the oldest stems every few years to rejuvenate the plant.

---

10 AUGUST

Purple gromwell *Glandora prostrata* 'Heavenly Blue'

PLANT TYPE: Evergreen shrub | **FAMILY:** Boraginaceae | **HEIGHT:** 15cm | **SPREAD:** 60cm | **EXPOSURE:** Full sun | **ASPECT:** South or west facing

The startling blue flowers of *Glandora prostrata* 'Heavenly Blue' will bring the front of a border to life. This spreading, mat-forming plant with evergreen foliage is perfect for groundcover. Its star-shaped flowers are low-growing, ideal for softening the edge of a border or for clothing a rockery, with blooms from May through to August. There are several beautiful varieties including 'Blue Star', which has blue flowers edged in white, or 'Alba', which has pure white flowers. Grow in well-drained, acidic soil in full sun. *Glandora prostrata* is sometimes sold as *Lithodora diffusa*.

11 AUGUST

Montbretia *Crocosmia* 'Paul's Best Yellow'

PLANT TYPE: Perennial | **FAMILY:** Iridaceae | **HEIGHT:** 1.5m | **SPREAD:** 1m | **EXPOSURE:** Full sun, Partial shade | **ASPECT:** South, east or west facing

Crocosmias add heat to late summer borders with flowers that come in shades of red, yellow and orange between July and September. The sprays of flowers are carried on arching spikes and look spectacular grown in drifts among plants such as grasses, rudbeckias and asters or combined with other hot-colour flowers to create an exotic look. 'Paul's Best Yellow' is a vigorous crocosmia with rich yellow flowers. The corms can be planted in spring, about 10cm deep in well-drained soil.

12 AUGUST

Madonna lily *Lilium candidum*

PLANT TYPE: Perennial | **FAMILY:** Liliaceae | **HEIGHT:** 1.5m | **SPREAD:** 50cm | **EXPOSURE:** Full sun | **ASPECT:** South, north, east or west facing

Among this plant's common names are Madonna lily and Annunciation lily, a reminder that this plant has long been associated with the Virgin Mary and is a symbol of purity for the Catholic Church. It has a long history and has been symbolic for many cultures, including the ancient Egyptians and the Greeks. This is a towering lily, at 1.5m tall, with trumpet-shaped fragrant white flowers. Plant bulbs in the autumn for flowers from June to August. The bulbs should be planted quite close to the surface of the soil, around 5–10cm deep. Stake plants before flowering time.

13 AUGUST

Dog rose *Rosa canina*

PLANT TYPE: Deciduous shrub | **FAMILY:** Rosaceae | **HEIGHT:** 4m | **SPREAD:** 3m | **EXPOSURE:** Full sun, Partial shade | **ASPECT:** South, north, east or west facing

The native dog rose is fast growing and prickly, ideal for a boundary hedge and a valuable wildlife plant. It attracts bees, butterflies and moths as well as providing food for birds such as blackbirds and redwings, and shelter for small mammals like hedgehogs. Its flowers are single and normally pink, appearing between May and August, followed by bright red hips that bring colour to the autumn garden. Prune older stems to around 30cm above ground level in winter to keep the shrub in good shape.

14 AUGUST

Night-scented stock *Matthiola longipetala*

PLANT TYPE: Annual | **FAMILY:** Brassicaceae | **HEIGHT:** 30cm | **SPREAD:** 30cm | **EXPOSURE:** Dappled shade, Full sun | **ASPECT:** South or west facing

Plant night-scented stock near your patio or garden bench, where you can enjoy the fragrance on a summer's evening. This hardy annual is a native plant, with lilac flowers that release their scent at night, making it a valuable plant for night-flying moths. It can be grown from seed by sowing direct outdoors in the spring, and it will flower between June and August. The flowers last for about three weeks, but a display can be extended by making sowings at three-week intervals.

---

15 AUGUST

Giant viper's bugloss *Echium pininana*

PLANT TYPE: Biennial | **FAMILY:** Boraginaceae | **HEIGHT:** 4m | **SPREAD:** 1m | **EXPOSURE:** Full sun | **ASPECT:** South or west facing

There's no missing this giant – *Echium pininana* is an impressive statement plant, with flower spikes that can reach 4m tall. A biennial plant from the Canary Islands, it is also known as Giant Viper's bugloss. In its first year it grows a rosette of silver leaves, then in the second year a huge flower spike appears with a mass of small, funnel-shaped blue flowers. It flowers from June to September. In warm regions of the UK, this plant will self-seed but the seed may not germinate in cold areas. Plant in a well-drained area with full sun for the best display.

16 AUGUST

Anderson's hebe *Veronica × andersonii*

PLANT TYPE: Evergreen shrub | **FAMILY:** Plantaginaceae | **HEIGHT:** 1m | **SPREAD:** 1m | **EXPOSURE:** Full sun, Partial shade | **ASPECT:** South, north, east or west facing

Hebes are useful shrubs with evergreen foliage and summer flowers that range from purple and pink to blue and white. They are mostly native to New Zealand, with some native to Australia and South America. Most hebes flower between June and September. Anderson's hebe grows to around 1m, although it can grow taller, with sprays of purple flowers that fade to white towards the end of flowering. Other attractive purple-flowered hebes include 'Autumn Glory' and 'Mohawk'. Plant in moist, well-drained soil and prune back after flowering to maintain the plant's shape.

17 AUGUST

Japanese anemone *Anemone × hybrida* 'Honorine Jobert'

PLANT TYPE: Perennial | **FAMILY:** Ranunculaceae | **HEIGHT:** 1.2m | **SPREAD:** 60cm | **EXPOSURE:** Full sun, Partial shade | **ASPECT:** South, north, east or west facing

'Honorine Jobert' is a top pick for a shady border, lighting up borders with its pure white flowers. It's in bloom from August to October, producing masses of cup-shaped, single flowers on tall stems. This elegant plant looks beautiful planted in among geraniums and grasses at the back of a border, or it can also be used for container displays as a foil for foliage plants such as hostas and ferns. Japanese anemones grow best in partial shade. To make new plants, divide clumps in early spring.

18 AUGUST

Red-hot poker *Kniphofia uvaria*

PLANT TYPE: Perennial | **FAMILY:** Asphodelaceae | **HEIGHT:** 1.2m | **SPREAD:** 60cm | **EXPOSURE:** Full sun, Partial shade | **ASPECT:** South or west facing

These dramatic perennials bring late-summer colour to borders with red and yellow flower spikes. At over a metre tall, these South African flowers, known as red-hot pokers because of their appearance, make spectacular planting partners for ornamental grasses, as well as other hot-coloured flowers like dahlias and crocosmias. The flowers appear from August to September and look best when planted in groups rather than as individual plants, although they do require a good amount of space. Plants can be divided in the autumn if they get too big.

19 AUGUST

Sunflower *Helianthus* 'Lemon Queen'

PLANT TYPE: Perennial | **FAMILY:** Asteraceae | **HEIGHT:** 2m | **SPREAD:** 45cm | **EXPOSURE:** Full sun | **ASPECT:** South, east or west facing

Although this is a type of sunflower, it is a perennial, unlike *Helianthus annuus*, which is an annual. 'Lemon Queen' has pale lemon-coloured, daisy-like flowers, with a darker yellow centre. It's a beautiful plant for the centre or back of a border, where it will produce a mass of flowers from July through to September. The flowerheads are 5cm across and the stems make good cut flowers to bring indoors – gather those that are still in bud for a longer flowering display. Insects also love these flowers, while birds will feed on the seedheads once flowering is over.

20 AUGUST

Feathertop *Pennisetum villosum*

PLANT TYPE: Perennial | **FAMILY:** Poaceae | **HEIGHT:** 60cm | **SPREAD:** 60cm | **EXPOSURE:** Full sun | **ASPECT:** South or west facing

The common names feathertop and fountain grass accurately describe this deciduous grass, which produces fluffy, feathery flowers between July and September. Native to north-eastern tropical parts of Africa, *Pennisetum villosum* is not hardy below −5°C and may not survive winter in cold areas of the UK. It can be protected with an autumn mulch or grown as an annual. It's an easy grass to grow either in a mixed border or for a summer container display as a contrast to vibrant flowers. Cut back top growth in early spring as new growth appears.

21 AUGUST

Tufted hair grass *Deschampsia cespitosa*

PLANT TYPE: Perennial | **FAMILY:** Poaceae | **HEIGHT:** 75cm | **SPREAD:** 75cm | **EXPOSURE:** Full sun, Partial shade | **ASPECT:** South, east or west facing

Deschampsia cespitosa is a native, evergreen grass that creates mounds of foliage. Over winter these provide valuable hibernation sites for insects such as ladybirds. The foliage is obscured over the summer, beneath a cloud of silvery purple flowerheads, at the top of tall stems. It's a spectacular grass to use in between shrubs and summer-flowering perennials such as achilleas and rudbeckias. By the end of summer, the flowers are dry and beige coloured, and often used in dried flower arrangements. This is an easy grass to grow, flourishing in sun or light shade.

22 AUGUST

Quaking grass *Briza media*

PLANT TYPE: Perennial | **FAMILY:** Poaceae | **HEIGHT:** 50cm | **SPREAD:** 50cm | **EXPOSURE:** Full sun | **ASPECT:** South, east or west facing

Briza media has some interesting common names: totter grass, dithery dock and quaking grass among them. They all relate to the heart-shaped flowerheads, which, as they are at the end of delicate stems, shiver or quake in the breeze. This grass grows in meadows and grasslands, but works well in ornamental borders, too, a foil for structural and colourful plants. The seeds provide valuable food for birds, such as linnets, greenfinches and house sparrows. Grow it in groups for maximum impact, in moist, well-drained soil. It can be cut back in spring if the foliage is looking dishevelled, or comb out any yellowing leaves.

23 AUGUST

Turkish sage *Phlomis russeliana*

PLANT TYPE: Perennial | **FAMILY:** Lamiaceae | **HEIGHT:** 90cm | **SPREAD:** 90cm | **EXPOSURE:** Full sun | **ASPECT:** South, east or west facing

Phlomis is native to Turkey and Syria, hence the common name, so it thrives best in full sun and can cope with hot, dry conditions. It's a pretty plant for herbaceous borders, with whorls of hooded yellow flowers up tall stems. Even after a long flowering period, from May through to September, it continues to add interest to the border with its dried stems and flowerheads, which look striking through winter, especially when layered in frost. Phlomis is a good plant for smothering weeds and looks good alongside grasses and perennials such as verbena. Other pretty varieties include the light pink-flowered 'Amazone'.

---

24 AUGUST

Ozark sundrops *Oenothera macrocarpa*

PLANT TYPE: Biennial | **FAMILY:** Onagraceae | **HEIGHT:** 20cm | **SPREAD:** 50cm | **EXPOSURE:** Full sun | **ASPECT:** South or east facing

Ozark sundrops, also known as evening primrose, is used in herbal remedies for ailments such as skin conditions and coughs. Several plants in this genus are referred to as evening primrose; the more well-known one, which is used for evening primose oil, is *Oenothera biennis*. It makes a pretty plant for growing over walls or at the edge of a border, where it will thrive in hot, dry conditions. The cup-shaped yellow flowers appear over a long period from July through to September, and being a spreading plant, it's a good choice for groundcover and suppressing weeds. Winged seedpods develop after the flowers, which are also attractive.

---

25 AUGUST

Small globe thistle *Echinops ritro*

PLANT TYPE: Perennial | **FAMILY:** Asteraceae | **HEIGHT:** 90cm | **SPREAD:** 45cm | **EXPOSURE:** Full sun, Partial shade | **ASPECT:** South, east or west facing

The globe thistle has striking blue flowers that pollinating insects love. Its round flowers contrast beautifully with the daisy-style, spires or flat-topped flowers of other summer-flowering perennials. *Echinops* is a wonderful plant for bringing architectural interest to plant displays and combines well with tall grasses like miscanthus or echinacea. It makes an excellent cut flower. 'Veitch's Blue' is a popular variety, with metallic-blue flowers and spikey leaves. Cut plants down after flowering to encourage a second flush of flowering. Lift and divide plants in autumn or spring if they become congested.

26 AUGUST

Musk mallow *Malva moschata*

PLANT TYPE: Perennial | **FAMILY:** Malvaceae | **HEIGHT:** 1m | **SPREAD:** 30cm | **EXPOSURE:** Full sun | **ASPECT:** South, east or west facing

The perennial wildflower musk mallow is native to southern England but is now found all over the UK. It's commonly seen alongside roads, in hedgerows and on the edges of fields, flourishing in dry soil and full sun. In gardens, this dainty flower is popular in cottage garden schemes and perennial wildflower meadows. For a white version, try 'Snow White', which is more compact at 45cm tall, and flowers until around September. Both types of musk mallow make excellent cut flowers and its musky fragrance is even easier to appreciate indoors.

27 AUGUST

Bluebeard *Caryopteris* × *cladonensis* 'Dark Knight'

PLANT TYPE: Deciduous shrub | **FAMILY:** Lamiaceae | **HEIGHT:** 1.2m | **SPREAD:** 1m | **EXPOSURE:** Full sun | **ASPECT:** South or west facing

At a time when many star players in the summer border are going over, *Caryopteris* × *cladonensis* is just getting into its stride. It's a beautiful upright shrub for late summer colour, with rich blue-purple flowers from August to September and silvery-green foliage. Thriving in hot, dry conditions, caryopteris is a good choice for those who want drought-tolerant or low-maintenance plants, whether in a border or a container, and also ideal for a wildlife-friendly garden as its flowers attract bees and butterflies. Prune back after flowering for a good display the following year.

28 AUGUST

Peruvian lily *Alstroemeria* 'Cahors'

PLANT TYPE: Perennial | **FAMILY:** Alstromeriaceae | **HEIGHT:** 80cm | **SPREAD:** 45cm | **EXPOSURE:** Full sun, Partial shade | **ASPECT:** South, east or west facing

Alstroemeria is also known as Peruvian lily or Lily of the Incas, because it's native to South America. It was brought to Europe by the Swedish botanist Baron von Alströmer in the 18th century, which is how it got its Latin name. Despite its origins and exotic appearance, alstroemeria are hardy here, with vibrant lily-like flowers that come in shades from red and orange through to softer shades such as pink and white. 'Cahors' has pink and yellow flowers over a long period from June to September.

---

29 AUGUST

Pink *Dianthus* 'Doris'

PLANT TYPE: Perennial | **FAMILY:** Caryophyllaceae | **HEIGHT:** 40cm | **SPREAD:** 60cm | **EXPOSURE:** Full sun | **ASPECT:** South or west facing

Pinks have traditionally been used in cottage gardens either at the front of borders or as part of a summer container display. They are named pinks not because so many of them come in shades of pink, but because of their serrated edges, which look like they have been cut with pinking shears. There are now many varieties to choose from, including modern types such as 'Doris', which has scented, double pale pink flowers with a darker centre and spreading silvery green foliage, which makes good groundcover. Deadhead regularly to keep the flowers coming and trim again in autumn to encourage new growth.

30 AUGUST

Mexican sunflower *Tithonia rotundifolia*

PLANT TYPE: Annual | **FAMILY:** Asteraceae | **HEIGHT:** 1.8m | **SPREAD:** 30cm | **EXPOSURE:** Full sun | **ASPECT:** South or west facing

The Mexican sunflower is a bright annual that adds height to borders at over a metre tall. Flowering from June, sometimes into autumn, it's useful for its hot colour and can be used for cut flowers. The variety 'Torch' is particularly vibrant and flowers until October – a good companion for other colourful and later-flowering plants such as dahlias and heleniums. Sow seed in mid to late spring indoors and plant out once the danger of frost has passed, as *Tithonia rotundifolia* is half hardy. It may need staking to prevent the plant flopping over.

31 AUGUST

Restharrow *Ononis spinosa*

PLANT TYPE: Perennial | **FAMILY:** Fabaceae | **HEIGHT:** 60cm | **SPREAD:** 60cm | **EXPOSURE:** Full sun | **ASPECT:** South or west facing

Also known as spiny restharrow, *Ononis spinosa* is native to the UK, where its natural habitat is chalky meadows. It has a shrubby looking appearance, with pea-like flowers that come in shades of pink, purple and blue. These nectar-rich blooms are attractive to bees and appear from May through to August. In the garden it's a lovely addition to rockeries, banks and slopes, especially if you're looking for good plants for pollinators. Grow in well-drained, neutral to acidic soil in a sunny spot.

SEPTEMBER

1 SEPTEMBER

Dahlia 'Topmix Pink' *Dahlia* 'Topmix Pink'

PLANT TYPE: Perennial | **FAMILY:** Asteraceae | **HEIGHT:** 30cm | **SPREAD:** 30cm | **EXPOSURE:** Full sun | **ASPECT:** South or west facing

The Topmix series of dahlias are low-growing, which makes them ideal for containers or at the front of borders. They have single flowers and a long flowering habit, with blooms brightening displays from June through to November. 'Topmix Pink' has pink flowers with a yellow centre but the series also has dahlias in shades including red, yellow and white. Dahlias grow best in full sun. Deadhead regularly to prolong flowering. In cold areas it may be necessary to lift the tubers after the first frost and store them over winter. In mild regions of the UK tubers can be left in the ground.

2 SEPTEMBER

Dahlia *Dahlia* 'Primrose Diane'

PLANT TYPE: Perennial | **FAMILY:** Asteraceae | **HEIGHT:** 1.2m | **SPREAD:** 30cm | **EXPOSURE:** Full sun | **ASPECT:** South or west facing

'Primrose Diane' is a decorative dahlia with small, pale yellow flowers but a tall habit, reaching over a metre tall. Decorative dahlias all have double flowers and the petals are usually broad and flat. They don't have a central disc and are the largest of all the dahlias – their flowers can be up to 25cm in diameter. Some other popular decorative dahlias include 'David Howard', which has orange-bronze flowers, perfect for warm displays, or 'Checkers', for something more unusual, with red and white, bi-colour petals and compact growth, reaching only 90cm tall.

3 SEPTEMBER

Dahlia *Dahlia* 'Gallery Art Nouveau'

PLANT TYPE: Perennial | **FAMILY:** Asteraceae | **HEIGHT:** 90cm | **SPREAD:** 90cm | **EXPOSURE:** Full sun | **ASPECT:** South or east facing

Dahlias are showy flowers with standout value in the border, like 'Gallery Art Nouveau', which has startingly cerise-pink flowers. It's a decorative dahlia with double flowers and is fairly low growing at 90cm tall. Dahlias are exotic on their own but provide even more impact when grown alongside contrasting or complementary plants. Some classic combinations include cosmos, nicotiana, *Verbena bonariensis*, achilleas and ornamental grasses such as *Pennisetum* 'Red Head' or *Stipa gigantea*. Dahlias are also ideal for use as cut flowers, providing as much impact indoors as out.

4 SEPTEMBER

Gazania *Gazania* 'Tiger Stripes'

PLANT TYPE: Perennial | **FAMILY:** Asteraceae | **HEIGHT:** 90cm | **SPREAD:** 60cm | **EXPOSURE:** Full sun | **ASPECT:** South or west facing

Gazanias are native to South Africa, where they grow in sunny hot conditions. In dull, grey weather, the flowers close up. When open, the plants provide a dramatic display with vibrant yellow and orange-striped daisy-like flowers that tie in with the common name 'Tiger Stripes'. Gazanias also come in shades of rose, red, pink, bronze, gold and orange with fresh green foliage. They are good container plants for a sunny patio or can be combined with other fiery colourful flowers for a hot-themed border. Deadhead regularly to keep the flowers coming.

5 SEPTEMBER

Hare's ear *Bupleurum rotundifolium*

PLANT TYPE: Annual | **FAMILY:** Apiaceae | **HEIGHT:** 90cm | **SPREAD:** 30cm | **EXPOSURE:** Full sun | **ASPECT:** South or west facing

Hare's ear is a native wild flower with acid-green flowers and foliage, popular with flower arrangers for the contrast it brings to more colourful plants. It looks especially attractive combined with white, red and blue-flowered plants. The yellow-green flowers look similar to euphorbia, with the true flowers surrounded by petal-like bracts. This plant has a long vase life and is used as a filler. It can be sown from seed, in spring or autumn. It flowers 12–14 weeks after sowing, so if sown in April, it should flower from August. Plants sown in autumn will grow taller.

6 SEPTEMBER

Yellow wax bells *Kirengeshoma palmata*

PLANT TYPE: Perennial | **FAMILY:** Hydrangeaceae | **HEIGHT:** 1.2m | **SPREAD:** 75cm | **EXPOSURE:** Full shade, Partial shade | **ASPECT:** North, east or west facing

Kirengeshoma palmata is prized for its foliage as much as its flowers, with maple-like leaves on an upright plant that's perfect for a shady spot. In late summer, it has clusters of waxy yellow flowers that are heavy enough to bend the striking maroon stems. These are shaped like fairy hats or bells, as the common name suggests. It grows best in partial shade, among other plants that relish these conditions, such as ferns, rodgersia, hostas and aconitums. Plant in a spot that provides shelter from the wind. It needs moist, lime-free soil. Dig in organic matter such as leafmould.

7 SEPTEMBER

Michaelmas daisy *Symphyotrichum* 'Little Carlow'

PLANT TYPE: Perennial | **FAMILY:** Asteraceae | **HEIGHT:** 90m | **SPREAD:** 45cm | **EXPOSURE:** Full sun, Partial shade | **ASPECT:** South, east or west facing

All asters used to be grouped under the same name, but some, including 'Little Carlow', are now grouped under the botanical name *Symphyotrichum*. Their common name Michaelmas daisy refers to the fact that their flowering coincides with the feast of Archangel Michael in late September. 'Little Carlow' has masses of pale purple flowers with a yellow centre and its height makes it ideal for the middle of borders, where it adds late colour to planting displays, as other flowering perennials may be starting to fade. It's also a bonus for late-flying insects on the hunt for nectar and pollen.

---

8 SEPTEMBER

Goldenrod *Solidago rugosa* 'Fireworks'

PLANT TYPE: Perennial | **FAMILY:** Asteraceae | **HEIGHT:** 1.5m | **SPREAD:** 70cm | **EXPOSURE:** Full sun | **ASPECT:** South, east or west facing

Goldenrods are herbaceous perennials, most of which are native to North America, where they grow in prairies and meadows. They were introduced to Britain in the 19th century, but became unpopular as some varieties can be invasive. 'Fireworks' is not one of these, and looks spectacular in a border growing alongside plants such as asters and *Verbena bonariensis* where its sprays of tiny yellow flowers complement the blues and purples of these flowers. Goldenrod is a useful plant for sandy soil as it grows well in these conditions. Deadhead regularly to avoid the plant self-seeding.

---

9 SEPTEMBER

Cuneate cinquefoil *Potentilla cuneata*

PLANT TYPE: Perennial | **FAMILY:** Rosaceae | **HEIGHT:** 10m | **SPREAD:** 10cm | **EXPOSURE:** Full sun | **ASPECT:** South, north, east or west facing

This low-growing perennial is native to the Himalayas and is well suited to growing in a rock garden. Its five-petalled, yellow flowers are in bloom over a long period from June into October. Its common name is cinquefoil, a name that in heraldry is an emblem of power and loyalty, and images of the five-petalled flower can be seen in buildings such as churches. This plant will grow well in most soils as long as it's not boggy and prefers a spot in full sun.

---

10 SEPTEMBER

Gaura *Oenothera lindheimeri* 'The Bride'

PLANT TYPE: Perennial | **FAMILY:** Onagraceae | **HEIGHT:** 90cm | **SPREAD:** 60cm | **EXPOSURE:** Full sun, Partial shade | **ASPECT:** South, east or west facing

For gardens that are short on space, having long-flowering plants in the border is invaluable and *Oenothera lindheimeri* 'The Bride' should be a top pick. Its delicate white flowers appear from May through to November, above soft foliage that creates good groundcover. It's well suited to cottage-garden-style planting but in its original habitat it grows in prairies, so would also make a good plant partner for ornamental grasses. Mulch in the autumn to protect the plant from cold winter weather.

11 SEPTEMBER

Bowden lily *Nerine bowdenii*

PLANT TYPE: Bulb | **FAMILY:** Amaryllidaceae | **HEIGHT:** 45cm | **SPREAD:** 10cm | **EXPOSURE:** Full sun | **ASPECT:** South, east or west facing

Nerines are native to South Africa and many are tender bulbs that need to be grown in a greenhouse. *Nerine bowdenii* is the hardiest nerine and can be grown outdoors in the UK in a warm sunny border. Plant this one in May or June for flowers from September to November. They are popular bulbs for their sprays of lily-like, bright pink flowers, which light up a herbaceous border but can also be used for cut flowers or grown in an autumn container display. Mulch well to protect them from winter cold, using a dry mulch such as bark.

12 SEPTEMBER

Bergamot *Monarda didyma* 'Cambridge Scarlet'

PLANT TYPE: Perennial | **FAMILY:** Lamiaceae | **HEIGHT:** 90cm | **SPREAD:** 45cm | **EXPOSURE:** Dappled shade, Full sun | **ASPECT:** South or west facing

One of monarda's common names is bee balm – although it is very attractive to bees, it has acquired this name because the plant was used to make a balm to soothe bee stings. *Monarda didyma* 'Cambridge Scarlet' has brilliant-red flowers, resembling a crazy hair-do, with curving flowers that grow from a central dome. If deadheaded regularly, monarda will flower for months. Bergamot foliage is fragrant and its leaves can be used for pot-pourri. Keep plants well watered to prevent powdery mildew.

13 SEPTEMBER

Honeywort *Cerinthe major* 'Purpurascens'

PLANT TYPE: Annual | **FAMILY:** Boraginaceae | **HEIGHT:** 60cm | **SPREAD:** 60cm | **EXPOSURE:** Dappled shade, Full sun | **ASPECT:** South or west facing

This annual plant has blue-green leaves and nodding, rich-purple bell-shaped flowers from May into late summer that are loved by bees. Although it's an annual, it will self-seed, so once you have grown plants, it should be easy to get more flowers for free the following year. Honeywort is a useful plant for filling gaps in the border and makes a beautiful cut flower because it has both striking foliage and flowers. Sow indoors from February or direct where they are to grow from April to May.

14 SEPTEMBER

Fuchsia *Fuchsia fulgens*

PLANT TYPE: Shrub | **FAMILY:** Onagraceae | **HEIGHT:** 1.1m | **SPREAD:** 1.1m | **EXPOSURE:** Full sun, Partial shade | **ASPECT:** South, east or west facing

Fuchsias are invaluable for their long-flowering nature, providing prolific blooms through summer and autumn, well into October. 'Mrs Popple' is a popular variety with large, vibrant red and purple flowers. Often grown as a feature plant in a border, it adds colour and height to planting displays. It can also be clipped and grown as a low-growing hedge. It not only attracts bees, butterflies and moths, but is also a food plant for caterpillars. Fuchsias put on the best display when planted in a sheltered spot, in fertile, moist but well-drained soil.

---

15 SEPTEMBER

Hardy plumbago *Ceratostigma plumbaginoides*

PLANT TYPE: Perennial | **FAMILY:** Plumbaginaceae | **HEIGHT:** 45cm | **SPREAD:** 30cm | **EXPOSURE:** Full sun | **ASPECT:** South or east facing

Hardy plumbago is grown for its clusters of bright blue flowers that bring late colour to borders from August to October. It has the bonus, though, of autumn leaf colour, with the bright green leaves turning a vivid red before falling. Even the red stems of this woody perennial add colour. Once established, *Ceratostigma plumbaginoides* is drought tolerant, which is useful for a sunny border as summers become hotter. It also has a spreading nature, making it a good pick for groundcover. Cut back any flowered shoots in spring.

16 SEPTEMBER

Japanese anemone *Anemone × hybrida* 'September Charm'

PLANT TYPE: Perennial | **FAMILY:** Ranunculaceae | **HEIGHT:** 90cm | **SPREAD:** 50cm | **EXPOSURE:** Full sun, Partial shade | **ASPECT:** South, north, east or west facing

Japanese anemones are elegant plants that will thrive in some shade, putting on a wonderful display in late summer and early autumn. The flowers are held on tall slender stems and this variety, 'September Charm', provides a profusion of soft pink flowers from July to September. Once established, it's a reliable plant that's ideal for growing in woodland locations and beneath trees or at the back of the border, where it can cope with dry conditions. Deadhead blooms after flowering and mulch annually with well-rotted manure or compost.

17 SEPTEMBER

Greater periwinkle *Vinca major*

PLANT TYPE: Shrub | **FAMILY:** Apocynaceae | **HEIGHT:** 45cm | **SPREAD:** 2.5m | **EXPOSURE:** Full shade, Partial shade, Full sun | **ASPECT:** South, north, east or west facing

Greater periwinkle is useful for covering stoney ground or a tricky bank with its pretty purple flowers, but it can become a problem in borders as it's extremely vigorous. It spreads by sending out long shoots that put down roots wherever they have contact with the ground. If you do have tricky areas that require groundcover, however, the violet flowers appear over a long period from April to September and will quickly add colour to areas beneath trees or sloping banks. For a border or small garden, lesser periwinkle is usually a better choice and a good way of suppressing weeds.

---

18 SEPTEMBER

Mexican cigar plant *Cuphea ignea*

PLANT TYPE: Perennial | **FAMILY:** Lythraceae | **HEIGHT:** 40cm | **SPREAD:** 40cm | **EXPOSURE:** Full sun, Partial shade | **ASPECT:** South or west facing

This tender perennial is also known as the firecracker plant because of its bright tubular flowers. They are a vibrant orange-red, with ash-coloured tips through the summer, hence its other common name, Mexican cigar plant. As it's tender, it should be grown as an annual outdoors or brought indoors for winter and grown as a houseplant. In either situation it's a vigorous grower with a prolific flowering habit between June and September. Sow indoors at a temperature of 18–20°C to germinate, then pot on when the seedlings are large enough to handle. Feed plants regularly through the growing season.

---

19 SEPTEMBER

Perennial phlox *Phlox paniculata*

PLANT TYPE: Perennial | **FAMILY:** Polemoniaceae | **HEIGHT:** 1.2m | **SPREAD:** 70cm | **EXPOSURE:** Full sun, Partial shade | **ASPECT:** South, east or west facing

Phlox paniculata is often called border phlox – it's the most common phlox to grow in the garden, with domes of small flowers. They are hardy plants that are easy to grow and a common feature of cottage-garden planting, although they can look more contemporary if paired with ornamental grasses. In the wild phlox can grow to almost 2m and has pink blooms, but there are now many compact cultivars to buy, in shades of blue, purple, pink, red, magenta and white. The flowers are often scented and attract a variety of bees and other pollinating insects.

---

20 SEPTEMBER

Anise scented sage *Salvia guaranitica* 'Black and Blue'

PLANT TYPE: Perennial | **FAMILY:** Lamiaceae | **HEIGHT:** 2.5m | **SPREAD:** 90cm | **EXPOSURE:** Full sun | **ASPECT:** South or west facing

'Black and Blue' is an appropriate name for this impressive plant, which has deep purple-blue flowers with almost black calcyces (the outer sepals which protect the flower). Flowering from August to October, on stems up to 2.5m tall, this is a stand-out plant for the back of a border. Its common name, anise-scented sage, refers to the aromatic foliage, which releases a scent when crushed or bruised. Plant in a sheltered position, south-facing if possible, as this plant is not completely hardy. Put down a layer of mulch around the plant in the autumn to protect the plant's roots from winter cold.

21 SEPTEMBER

Seven son flower tree *Heptacodium miconioides*

PLANT TYPE: Shrub | **FAMILY:** Caprifoliaceae | **HEIGHT:** 6m | **SPREAD:** 3m | **EXPOSURE:** Full sun, Partial shade | **ASPECT:** South, east or west facing

This hardy shrub is native to China but is fully hardy in the UK. It can reach 6m tall, and provides year-round interest – it has glossy foliage through summer, then white scented flowers appear in late summer and last until around November. Once the flowers fall, there is still colour from the pink-red calyxes, which turn red-purple over time. In winter, the bark becomes a feature as the light outer bark peels off leaving a dark-coloured shade underneath. This shrub is easy to grow and just needs some shelter from cold winds.

22 SEPTEMBER

Wax plant *Hoya carnosa*

PLANT TYPE: Evergreen climber | **FAMILY:** Apocynaceae | **HEIGHT:** 4–8m | **SPREAD:** 1–1.5m | **EXPOSURE:** Full sun | **ASPECT:** South or west facing

The tender wax flower is part of a group of plants native to Asia, Australia and the Pacific Islands. It needs a minimum temperature of 10 degrees Celsius to flourish, so it is usually grown in containers and brought indoors when temperatures drop or used as an exotic house plant. Its star-shaped, waxy white flowers are night scented and borne in clusters on climbing stems. *Hoya carnosa* needs humidity – if grown indoors, spray regularly to increase humidity. Feed monthly during the growing season. Outdoors, this climber can grow to 6m or more, but indoors it's unlikely to grow beyond 3m. Prune to keep it to a manageable size.

23 SEPTEMBER

Zebra grass *Miscanthus sinensis* 'Zebrinus'

PLANT TYPE: Perennial | **FAMILY:** Poaceae | **HEIGHT:** 1.25m | **SPREAD:** 45cm | **EXPOSURE:** Full sun | **ASPECT:** South, north, east or west facing

It's the cream bands on this grass's stems that give it the name Zebra grass. These markings in themselves give the plant impact, but it also has silky flowers that appear in August and September. These can last for months and the grass is a worthy addition to the border for adding structure through autumn and winter. At over a metre tall, it is a good pick for the back of a border where it can act as a foil for more colourful plants. Cut back to the ground in late winter to early spring before the new growth emerges.

24 SEPTEMBER

Coneflower *Rudbeckia fulgida* var. *sullivantii* 'Goldsturm'

PLANT TYPE: Perennial | **FAMILY:** Asteraceae | **HEIGHT:** 60cm | **SPREAD:** 45cm | **EXPOSURE:** Full sun, Partial shade | **ASPECT:** South, east or west facing

Rudbeckia fulgida var. *sullivantii* 'Goldsturm' is a reliable, late-flowering perennial that is also known as coneflower or black-eyed Susan. It's easy to grow because it's compact and doesn't need staking and it doesn't spread too fast, making it easy to maintain. Grown in drifts, in between other perennials or ornamental grasses, it lights up autumn borders until October. Some classic combinations for this golden daisy flower include perovskia, echinacea and persicaria. For the best display, grow rudbeckias in moist, well-drained oil in full sun.

25 SEPTEMBER

Common bistort *Bistorta officinalis* 'Superba'

PLANT TYPE: Perennial | **FAMILY:** Polygonaceae | **HEIGHT:** 90cm | **SPREAD:** 90cm | **EXPOSURE:** Full sun, Partial shade | **ASPECT:** South, east or west facing

This is a plant that is worth planting in generous amounts, where it can make an impact and has room to spread. The spire-shaped flowers are light pink and appear from June to September above a carpet of rich green foliage that takes on autumn colour. It's a fast spreader, perfect for using as groundcover in part shade or sun near the front of the border. Once planted, it is low maintenance and complements plants such as achilleas, asters and dahlias. Grow it in moist, well-drained soil and cut it back after flowering. Divide congested clumps every three years.

26 SEPTEMBER

Glossy abelia *Abelia × grandiflora*

PLANT TYPE: Shrub | **FAMILY:** Caprifoliaceae | **HEIGHT:** 3m | **SPREAD:** 4m | **EXPOSURE:** Full sun | **ASPECT:** South or east facing

The glossy part of this plant's common name refers to the lush green leaves, which are semi-evergreen and look attractive for much of the year. *Abelia × grandiflora* makes a lovely feature shrub at the back of the border, with clusters of small, pale pink flowers that appear for months, right up to October. This shrub benefits from a warm, sunny spot, where its fragrance can be enjoyed. It can also be clipped into an attractive hedge. In autumn, trim back the flowered shoot after flowering to keep the plant tidy.

27 SEPTEMBER

Cup and saucer vine *Cobaea scandens*

PLANT TYPE: Perennial climber | **FAMILY:** Polemoniaceae | **HEIGHT:** 4m | **SPREAD:** 1m | **EXPOSURE:** Full sun | **ASPECT:** South or west facing

Cobaea scandens is a fast-growing tropical plant with exotic purple flowers up to 8cm long. It can reach up to 6–7m unpruned but in the UK grows to around 4m, ideal for clothing pergolas or trellis in colourful summer flowers which release a sweet fragrance. It won't survive British winters as it needs a temperature above 7 degrees Celsius and is therefore best grown as an annual or as a container plant, which can be moved indoors in the autumn. Feed with an organic tomato feed once a fortnight to keep the flowers coming and deadhead regularly.

28 SEPTEMBER

Hyssop *Hyssopus officinalis*

PLANT TYPE: Perennial | **FAMILY:** Lamiaceae | **HEIGHT:** 75cm | **SPREAD:** 90cm | **EXPOSURE:** Full sun, Partial shade | **ASPECT:** South or west facing

Hyssop is a strongly flavoured Mediterranean herb, which was used in the Middle Ages to boost soups and stews and is similar in looks to rosemary and lavender. These days, it's mainly used to flavour liqueurs such as Chartreuse. The flower spikes are a violet blue to deep blue and attractive to bees and other pollinators. As a herb and ornamental plant, it will work well in a herb garden or herbaceous border, as well as containers. Its flowers are in bloom from July to September. Grow it in moist, well-drained soil.

29 SEPTEMBER

Speedwell *Veronica longifolia* 'Marietta'

PLANT TYPE: Perennial | **FAMILY:** Plantaginaceae | **HEIGHT:** 65cm | **SPREAD:** 45cm | **EXPOSURE:** Full sun, Partial shade | **ASPECT:** South, north, east or west facing

Speedwell is useful for adding a vertical accent to planting displays, with its long spires of flowers. 'Marietta' is a striking cultivar, more compact than the species, with indigo-coloured flowers over a long period from June to October. For small gardens, this is a good investment as it brings colour to the border for months on end as well as attracting pollinating insects. This plant may need staking in windy locations. It grows best in moist, well-drained soil. Deadhead regularly to keep this pretty plant flowering into autumn.

---

30 SEPTEMBER

Summer hyacinth *Galtonia candicans*

PLANT TYPE: Bulb | **FAMILY:** Asparagaceae | **HEIGHT:** 1.2m | **SPREAD:** 40cm | **EXPOSURE:** Full sun | **ASPECT:** South, east or west facing

Waxy white flowers dangle from tall stems around a metre tall on this bulbous perennial. *Galtonia candicans* look good planted in drifts and can also be cut for indoor displays where it may be easier to appreciate the delicate fragrance of their flowers. Also known as summer hyacinth, this plant flowers from July to October. Plant the bulbs from March to June, 30cm apart and around 10–15cm deep in a sunny position. They need a warm spot as they are native to South Africa. In areas with cold winters, it is best to lift the bulbs and store them in a frost-free place.

OCTOBER

1 OCTOBER

Autumn crocus *Colchicum* 'Waterlily'

PLANT TYPE: Bulb | **FAMILY:** Colchicaceae | **HEIGHT:** 10cm | **SPREAD:** 10cm | **EXPOSURE:** Full sun | **ASPECT:** South, north, east or west facing

Autumn crocus are not related to true crocuses, despite their common name. There are both spring- and autumn-flowering *Colchicum* species, with the most common types flowering in September and October. The autumn crocus 'Waterlily' has double pink to purple flowers that are shaped like a waterlily. The flowers of autumn crocus appear without leaves, hence their other common name, 'naked ladies'. These attractive flowers will thrive in a sunny border or naturalise well in grass. Plant the bulbs in August, 5–10cm deep and 15cm apart.

2 OCTOBER

Chrysanthemum *Chrysanthemum* 'Green Mist'

PLANT TYPE: Perennial | **FAMILY:** Asteraceae | **HEIGHT:** 50cm | **SPREAD:** 50cm | **EXPOSURE:** Full sun | **ASPECT:** South or west facing

Chrysanthemums are valued for their late flowers, which brighten September and October borders with their bold blooms. There is a huge range of colours to choose from. 'Green Mist', with its spikey lime-green flowers, adds a contemporary look to displays, but for hot schemes there are many vibrant orange, yellow and red types to choose from, which also come in a variety of flowerhead shapes, from pompom to single. Grow these dramatic flowers in well-drained soil, and stake the plants if the site is windy or exposed.

3 OCTOBER

Mexican giant hyssop *Agastache* 'Blue Fortune'

PLANT TYPE: Perennial | **FAMILY:** Lamiaceae | **HEIGHT:** 1m | **SPREAD:** 40cm | **EXPOSURE:** Full sun | **ASPECT:** South or west facing

Drought-tolerant plants like *Agastache* are ideal for flowerbeds that get sun for much of the day. 'Blue Fortune' is easy to look after and long flowering, with blue bottlebrush flowers from June through to October. This plant provides a striking vertical accent in planting displays, a good contrast for spreading plants like hardy geraniums or daisy shaped flowers, such as heleniums. They also have aromatic foliage that has a liquorice fragrance and the flowers are extremely attractive to pollinators. Plant in a sunny border that provides some shelter from cold winds. It can be short-lived.

4 OCTOBER

Abyssinian gladiolus *Gladiolus murielae*

PLANT TYPE: Bulb | **FAMILY:** Iridaceae | **HEIGHT:** 1m | **SPREAD:** 40cm | **EXPOSURE:** Full sun | **ASPECT:** South, east or west facing

Hailing from East Africa, this elegant gladiolus thrives in full sun. Planted in the spring, these bulbs will flower from August to October, providing late interest with their star-shaped white flowers that have a maroon centre. These blooms are held at the top of slender stems and look best when planted in a group or drift, adding movement and height to late-summer displays. Stake plants before the flowers spike appears as these bulbs can grow to a metre tall. Although this bulb is a perennial, it's best treated like an annual, and replaced each year, similar to tulips.

---

5 OCTOBER

Begonia *Begonia* 'Illumination Scarlet'

PLANT TYPE: Perennial | **FAMILY:** Begoniaceae | **HEIGHT:** 30cm | **SPREAD:** 30cm | **EXPOSURE:** Partial shade | **ASPECT:** North, east or west facing

For a mass of summer flowers, begonias are a winning choice. They come in bright colours, from pink and yellow, through to orange, red and white. Tuberous begonias like 'Illumination Scarlet' flower through summer and into autumn, making excellent container plants. They are usually used as annuals, as they don't survive winter, although the tubers can be brought indoors and stored. This variety has large, bright red flowers from June to October on a cascading plant that is useful for softening the edges of container displays and hanging baskets.

6 OCTOBER

Autumn daffodil *Sternbergia lutea*

PLANT TYPE: Bulb | **FAMILY:** Amaryllidaceae | **HEIGHT:** 15cm | **SPREAD:** 10cm | **EXPOSURE:** Full sun | **ASPECT:** South, east or west facing

Common names for *Sternbergia lutea* include lily of the field and yellow starflower, referring mainly to its appearance, although its yellow goblet-shaped flowers look more like crocus than lily. *Sternbergia* is in honour of the botanist Count Kaspar Moritz von Sternberg, and the *lutea* of its Latin name simply means yellow. It's native to the Mediterranean and thrives in full sun, flowering from September to October. It looks most striking when planted in drifts or en masse. Plant bulbs August to September; if the bulbs are planted between October and December, they will flower the following year.

7 OCTOBER

Shrubby cinquefoil *Potentilla fruticosa* 'Elizabeth'

PLANT TYPE: Shrub | **FAMILY:** Rosaceae | **HEIGHT:** 1m | **SPREAD:** 1m | **EXPOSURE:** Full sun, Partial shade | **ASPECT:** South, north, east or west facing

Potentilla fruticosa 'Elizabeth' is also known as the buttercup shrub because of its cheery yellow flowers. It's a prolific flowering shrub, adding colour to borders from spring through to autumn. At a metre tall, its compact size makes it ideal for the middle or front of a border, where it will provide interest over a long period. Cinquefoil are deciduous shrubs that are useful for mixed borders, or as low hedging. As well as yellow there are also white, orange and pink flowering varieties. Trim plants annually before the flower buds develop to keep them neat.

---

8 OCTOBER

Heath aster *Symphyotrichum ericoides* 'Pink Cloud'

PLANT TYPE: Perennial | **FAMILY:** Asteraceae | **HEIGHT:** 60cm | **SPREAD:** 45cm | **EXPOSURE:** Full sun | **ASPECT:** South, east or west facing

This aster is tougher than many, a good choice for coastal areas and beds with dry soil. It forms attractive mounds covered in small lilac-pink flowers with yellow centres. This late-flowering perennial blooms from August to October. Although the flowers are small, the plant produces such a number of them that it creates a fantastic display as other plants are fading. It goes well with ornamental grasses, dahlias or late-flowering perennials such as sanguisorbas or rudbeckias. Plant it in a well-drained position, in a sunny spot.

---

9 OCTOBER

Crimson flag *Hesperantha coccinea* 'Jennifer'

PLANT TYPE: Perennial | **FAMILY:** Iridaceae | **HEIGHT:** 60cm | **SPREAD:** 30cm | **EXPOSURE:** Full sun | **ASPECT:** South or east facing

Crimson flag is in the iris family and has similar characteristics, with its sword-shaped leaves, previously known under the name *Schizostylis*. *Hesperantha coccinea* is a wonderful autumn-flowering plant that reaches its peak as many plants are dying down. They look a little like gladioli with upright flower spikes, each of which have several flowers. 'Jennifer' has soft pink flowers, cup-shaped like crocus, which appear from August to November. Plant *Hesperantha* in well-drained soil, it looks lovely alongside grasses or other late-flowering perennials such as asters.

---

10 OCTOBER

Big blue lilyturf *Liriope muscari*

PLANT TYPE: Perennial | **FAMILY:** Asparagaceae | **HEIGHT:** 30cm | **SPREAD:** 45cm | **EXPOSURE:** Full shade, Partial shade | **ASPECT:** North, east or west facing

Lilyturf is not only low maintenance but it's also a valuable plant for dry, shady areas. It can be grown beneath trees and shrubs, where it will create easy ground cover for tricky spots, producing lots of blue-purple flower spikes in autumn. It can go on flowering into November and has evergreen, grass-like foliage. Reaching around 30cm tall, it's a good plant for the front of a border or planting into gaps between other shade lovers. Grow it in well-drained soil in partial to full shade. If clumps get congested, divide in the spring.

11 OCTOBER

Spanish flag *Ipomoea lobata*

PLANT TYPE: Annual | **FAMILY:** Convolvulaceae | **HEIGHT:** 3m | **SPREAD:** 1m | **EXPOSURE:** Full sun | **ASPECT:** South or west facing

Spanish flag has the same colours as the country's real flag, red and yellow, arranged along long 30cm stems. The red flowers fade to yellow then cream. As the flowers are aging at different times, there's always a mix of vibrant colours to brighten a container or border. This fast-growing annual climber is perfect for covering wigwams in containers or trellis, bringing bright colour to gardens and patios from summer into autumn. Sow seed indoors from March to May and plant out once the last frosts are over. Plant in a sunny spot. Keep deadheading to encourage more flowers.

12 OCTOBER

Trailing abutilon *Abutilon megapotamicum*

PLANT TYPE: Evergreen shrub | **FAMILY:** Malvaceae | **HEIGHT:** 2m | **SPREAD:** 2m | **EXPOSURE:** Full sun | **ASPECT:** South or west facing

Abutilon megapotamicum is a shrub native to the mountain valleys of Brazil, ideal for growing in a cool conservatory or in a container on the patio. This elegant shrub has lantern-shaped flowers over a long period, from April to October, hanging from slender stems. There are many cultivars to choose from in colours ranging from white to pale yellow, pink and deep red. They were popular in Victorian times as bedding plants, providing bright colour in planting displays. As this shrub is half hardy it will need to be brought in over winter. In mild regions of the UK it could be grown against a warm wall if sheltered from wind and frost.

13 OCTOBER

Lion's tail *Leonotis leonurus*

PLANT TYPE: Semi-evergreen shrub | **FAMILY:** Lamiaceae | **HEIGHT:** 2m | **SPREAD:** 1m | **EXPOSURE:** Full sun | **ASPECT:** South or west facing

The common name for *Leonotis leonurus* is Lion's ear or Lion's tail, perhaps because of the shape of the orange flowers. The shrub comes from South Africa, growing in grassland and used traditionally as a treatment for snake bites as well as a remedy for fevers and coughs. In the UK it is often treated as a tender perennial, and needs a sheltered position in full sun to flourish, or it can be grown in a cool greenhouse. The whorls of rich orange-red flowers are carried on upright stems, up to 2m tall.

14 OCTOBER

Amarine × *Amarine* 'Anastasia' (Belladiva Series)

PLANT TYPE: Bulb | **FAMILY:** Amaryllidaceae | **HEIGHT:** 50cm | **SPREAD:** 50cm | **EXPOSURE:** Full sun | **ASPECT:** South or west facing

Amarines are the result of a cross between *Nerine bowdenii* and *Amaryllis belladonna* – a plant with larger flowers, more of them and a stronger growing habit. This autumn-flowering bulb can be planted in May to June, for flowers between September and November. It's easy to grow, reliable and the showstopping colour makes an exotic display in the border. 'Anastasia' is a bright pink variety that looks striking planted among dahlias and other autumn-flowering plants such as chrysanthemums. It is also an excellent cut flower.

---

15 OCTOBER

Borage *Borago officinalis*

PLANT TYPE: Annual | **FAMILY:** Boraginaceae | **HEIGHT:** 60cm | **SPREAD:** 45cm | **EXPOSURE:** Full sun | **ASPECT:** South or west facing

Borage is annual herb with cucumber-flavoured leaves and edible flowers. The leaves and flowers can be added to drinks and salads and have been used over the centuries in concoctions to inspire romance, courage in soldiers from the Celts to the Greeks, and to drive away sorrow. It's attractive to bees and other pollinating insects, making it a useful companion plant in the vegetable patch. At the end of the season it can be added to the compost heap, where it will provide bulk and added nutrients. Sow this herb in spring for flowers from June through to October.

16 OCTOBER

Dahlia *Dahlia* 'Frigoulet'

PLANT TYPE: Perennial | **FAMILY:** Asteraceae | **HEIGHT:** 1.4m | **SPREAD:** 40cm | **EXPOSURE:** Full sun | **ASPECT:** South or west facing

Semi-cactus dahlias have star-like flowers, which means they cope well in bad weather, as the flowers are less dense than other types and don't get weighed down by heavy rain. They have fully double flowers but the petals have a broader base than cactus dahlias and are rolled in until about halfway along.
A spectacular example is 'Figoulet', which has bi-coloured flowers, a mix of red at the base with white tips. Not only does it stand out in the border but it also makes a beautiful cut flower.

17 OCTOBER

Ice plant *Delosperma cooperi* 'Jewel of Desert Peridott'

PLANT TYPE: Perennial | **FAMILY:** Aizoaceae | **HEIGHT:** 15cm | **SPREAD:** 20cm | **EXPOSURE:** Full sun | **ASPECT:** West facing

Delosperma are small, succulent plants native to southern and eastern Africa, where they grow on mountains and come in a spectrum of colours including white, orange, yellow, red, pink and purple. 'Jewel of Desert Peridott' is part of the Jewel of Desert Series, with yellow, daisy like flowers that appear over a long period from April to October. These plants are perfect for rockeries and areas of poor soil, where they will spread to create a carpet of colour. It's also a good choice for underplanting in containers.

---

18 OCTOBER

Strawberry tree *Arbutus unedo*

PLANT TYPE: Evergreen shrub | **FAMILY:** Ericaceae | **HEIGHT:** 8m | **SPREAD:** 8m | **EXPOSURE:** Full sun | **ASPECT:** South or west facing

The strawberry tree has bell-shaped autumn flowers, which are white sometimes with a hint of pink, and these appear at the same time as the previous year's fruit is ripening. The fruit looks like strawberries, which is where the common name comes from, but they do not taste good when eaten raw. The species name is thought to come from the Latin *unem edo*, 'I eat one', due to the unpleasant taste when fresh, although they are used in some countries to make jam or liqueurs. Plant in a sheltered spot in well-drained soil. It will grow well in coastal areas as well as inland.

---

19 OCTOBER

Red clover *Trifolium pratense*

PLANT TYPE: Perennial | **FAMILY:** Fabaceae | **HEIGHT:** 30cm | **SPREAD:** 40cm | **EXPOSURE:** Full sun | **ASPECT:** East, south or west facing

Red clover is a native perennial wildflower with attractive pink-red flowers that are a magnet for bees and other pollinating insects. It can be seen growing in wildflower meadows and grasslands and is often used as a green manure, as it fixes nitrogen in the soil. It was traditionally added to hay meadows because of its soil-improving properties. It's a great addition to a mini meadow or even for filling gaps in a border, where it will thrive in well-drained soil. Sow from March to May for flowers from early summer through to October, or buy as plug plants to save time.

---

20 OCTOBER

Red-hot poker *Kniphofia* 'Ice Queen'

PLANT TYPE: Perennial | **FAMILY:** Asphodelaceae | **HEIGHT:** 1.2m | **SPREAD:** 75cm | **EXPOSURE:** Full sun, Partial shade | **ASPECT:** South or west facing

Red-hot pokers always make a statement in the border, with their fiery red and orange flowers. For something different, the variety 'Ice Queen' has pale, green-tipped white flower spikes that would fit well in a cooler planting scheme or as a foil for hot-coloured plants such as heleniums and rudbeckias or plants that enjoy similar growing conditions, such as the shrub *Euphorbia mellifera* with its attractive foliage. Like other red-hot pokers it has a long flowering season, from July through to October.

21 OCTOBER

Baltic parsley *Cenolophium denudatum*

PLANT TYPE: Perennial | **FAMILY:** Apiaceae | **HEIGHT:** 1.5m | **SPREAD:** 75–90cm | **EXPOSURE:** Full sun, Partial shade | **ASPECT:** South, east or west facing

Baltic parsley is not edible, but is grown for its ornamental qualities, with domed flowerheads that attract insects. It has become fashionable in the UK in wild gardens and nature-friendly borders with flowers from July to October. Even after this it provides architectural value through winter with attractive seedheads that are enhanced by a covering of frost and can be used in cut flower or dried flower arrangments. When hard frosts arrive, the plant will die back and regrow the following spring. It's hardy down to −10°C.

22 OCTOBER

Rose campion *Lychnis coronaria*

PLANT TYPE: Perennial | **FAMILY:** Caryophyllaceae | **HEIGHT:** 1m | **SPREAD:** 50cm | **EXPOSURE:** Full sun | **ASPECT:** South or west facing

Rose campion flowers for months and is easy to grow from seed. Although it's a short-lived perennial, it self-seeds, producing new plants for free that can be grown on to fill any gaps. To avoid self-seeders, remove the spent flowers before they form seedheads. The deep pink flowers are held on tall, branching stems, contrasting well with its silver foliage. It creates a cottage-garden-style display when combined with perennials such as verbascum and hardy geraniums, or for something more contemporary, grow it alongside ornamental grasses. Plant it in a sunny spot – once established it is drought tolerant.

23 OCTOBER

Balkan spurge *Euphorbia oblongata*

PLANT TYPE: Perennial | **FAMILY:** Euphorbiaceae | **HEIGHT:** 50m | **SPREAD:** 40cm | **EXPOSURE:** Partial shade | **ASPECT:** North, east or west facing

Although this plant is a perennial, it's usually short-lived at its flowering best in the first year. For this reason, it's easier to treat it as an annual. When sown in the spring it flowers for months and months, from May to December. Despite its long flowering time, it's a popular foliage plant in cut-flower arrangements, acting as a foil for more colourful flowers. Balkan spurge is also useful for filling gaps in borders or container displays, its vivid green colours complementing shades such as purple and orange. It's easy to grow and low-maintenance too.

---

24 OCTOBER

Sensitive plant *Mimosa pudica*

PLANT TYPE: Annual | **FAMILY:** Fabaceae | **HEIGHT:** 30cm | **SPREAD:** 25cm | **EXPOSURE:** Full sun, Partial shade | **ASPECT:** South or west facing

Mimosa pudica is named sensitive plant because its leaves fold up and its stems droop if they are touched. The leaves also fold up at night. Some of its other common names include touch-me-not and sensible plant. Although it's a short-lived perennial, it deteriorates with age, so it is more often grown as an annual house plant. It's easy to grow from seed, and will produce pink to purple flowers from July to October. These are followed by seed pods, from which the seed can be saved for more plants the following year. Grow it in a humid spot, in a warm room (above 18°C).

---

25 OCTOBER

Tree hollyhock *Hibiscus syriacus* 'Woodbridge'

PLANT TYPE: Deciduous shrub | **FAMILY:** Malvaceae | **HEIGHT:** 2m | **SPREAD:** 2m | **EXPOSURE:** Full sun | **ASPECT:** South or west facing

Also known as Rose of Sharon, *Hibiscus syriacus* is the national flower of Korea. Rose of Sharon represents beauty in the bible, the name perhaps in tribute to the trumpet-shaped flowers with a red centre. This species has cultivars with flowers in shades of pink, white or purple. 'Woodbridge' has rich pink flowers with a red centre, blooming late in the season from August to October. The dark green foliage is also attractive, although it doesn't appear until late spring. Plant this shrub in a sunny border for the best display of flowers.

26 OCTOBER

St John's wort *Hypericum × hidcoteense* 'Hidcote'

PLANT TYPE: Evergreen/semi-evergreen shrub | **FAMILY:** Hypericaceae | **HEIGHT:** 1.2m | **SPREAD:** 1.5m | **EXPOSURE:** Full sun, Partial shade | **ASPECT:** South, east or west facing

Cup-shaped yellow flowers cover this shrub from July to October in a display that will enliven a border as early summer flowers fade. The flowers are followed by red-brown berries. It has a dense habit and not only makes an attractive and reliable shrub but can also be grown as a low hedge or to create screening. St John's wort is well known as a herbal treatment for depression, but that particular St John's wort is *Hypericum perforatum*. This species is a shrubby hypericum, which should be pruned in spring to keep it tidy. Plant in a sheltered position if possible.

27 OCTOBER

Mexican fleabane *Erigeron karvinskianus*

PLANT TYPE: Perennial | **FAMILY:** Asteraceae | **HEIGHT:** 30cm | **SPREAD:** 1m | **EXPOSURE:** Full sun | **ASPECT:** South, east or west facing

This is a versatile perennial for many situations, perfect for filling gaps in containers, softening walls or edging borders. The daisies are white, then turn pink as they age, which means the plant always has two colours as more flowers appear over the summer. It's also incredibly long-flowering, lasting from May to October. Unsurprisingly, given its common name, it's native to Mexico and Central America, but it thrives in British gardens when planted in sunny spots and is a useful drought-tolerant plant for dry areas. They are also a wonderful wildlife plant, attracting bees and butterflies to the garden.

28 OCTOBER

Silver banner grass *Miscanthus sacchariflorus*

PLANT TYPE: Deciduous grass | **FAMILY:** Poaceae | **HEIGHT:** 3m | **SPREAD:** 1.4m | **EXPOSURE:** Full sun, Partial shade | **ASPECT:** South, north, east or west facing

In a large garden, this tall grass is good for creating shelter from the wind, or for use as screening. It has bamboo-like stems with long, arching leaves that have a white midrib. The fluffy plumes of flowers appear from late summer into October. *Miscanthus sacchariflorus* flowers most reliably after a hot summer or in mild regions and these remain a feature throughout winter, making it a good focal plant for several seasons. As this is a deciduous grass, it should be cut down to the ground in late February to March.

29 OCTOBER

Red valerian *Centranthus ruber*

PLANT TYPE: Perennial | **FAMILY:** Caprifoliaceae | **HEIGHT:** 1m | **SPREAD:** 50cm | **EXPOSURE:** Full sun | **ASPECT:** South or west facing

Red valerian is a perennial wallflower with fragrant pink, crimson or white flowers on tall stems. It's a cottage-garden-style plant that looks good planted in groups. It will grow well in a border alongside plants such as scabious, pinks and gypsophila, but is at its best on stony ground and is a good choice for slopes. Sometimes it self-seeds into old walls and it suits this haphazard way of growing. Red valerian is long flowering, blooming from July into October. Sow the seed in early spring.

30 OCTOBER

Brazilian fuchsia *Justicia floribunda rizzinii*

PLANT TYPE: Evergreen shrub | **FAMILY:** Acanthaceae | **HEIGHT:** 60cm | **SPREAD:** 60cm | **EXPOSURE:** Partial shade | **ASPECT:** South or east facing

This is a frost-tender shrub that needs to be grown in a greenhouse, or in a mild region where it should be brought indoors over winter. In its native habitat in Eastern Brazil, it grows in subtropical forests and the tubular yellow and red flowers attract hummingbirds who drink the nectar and pollinate the flowers. It is a valuable conservatory or greenhouse plant as it flowers from late autumn into early spring, the exotic blooms set off by evergreen foliage. It will grow best with a minimum winter temperature of 13°C.

31 OCTOBER

Sweet tea *Osmanthus fragrans* f. *aurantiacus*

PLANT TYPE: Evergreen shrub | **FAMILY:** Oleaceae | **HEIGHT:** 6m | **SPREAD:** 6m | **EXPOSURE:** Full sun | **ASPECT:** South or west facing

Known as sweet tea or fragrant olive, *Osmanthus fragrans* f. *aurantiacus* is a tender shrub with scented orange flowers. It's autumn flowering, with blooms in September and October. The flowers are followed by black berries. Occasionally it will also flower in spring or summer. It is native to China and Japan, and although it can be grown next to a warm wall in mild regions, it is only frost hardy down to −5°C, so it might be better suited to a conservatory or unheated greenhouse.

NOVEMBER

1 NOVEMBER

Fortune saxifrage *Saxifraga fortunei* 'Silver Velvet'

PLANT TYPE: Perennial | **FAMILY:** Saxifragaceae | **HEIGHT:** 20cm | **SPREAD:** 30cm | **EXPOSURE:** Full shade, Partial shade | **ASPECT:** North or east facing

Many saxifrages are suited to rock gardens, being low-growing and native to mountainous regions in China and Japan. *Saxifraga fortunei* was named after the Scottish plant hunter, Robert Fortune, who discovered it in China. It grows well in rock gardens or in leafy, moist soil, flowering in autumn and dying back when the hard frosts arrive. 'Silver Velvet' is a pretty variety with white flowers from September to November and neat mounds of scalloped, burgundy leaves that have grey stripes. It is perfect for brightening a shady spot in the garden, whether in a border or a container.

2 NOVEMBER

Oregon grape *Mahonia nitens* 'Cabaret'

PLANT TYPE: Evergreen shrub | **FAMILY:** Berberidaceae | **HEIGHT:** 1m | **SPREAD:** 80cm | **EXPOSURE:** Full sun, Partial shade | **ASPECT:** South, north, east or west facing

'Cabaret' is attractive even before it blooms, with fiery orange flower buds that open into deep-yellow flowers. This vibrant shrub will add colour to gardens from August into winter, the flowers lasting until November, followed by attractive blue berries. The flower spikes are long, but the plant itself is compact, reaching only 1m in height, compared to the 4m or more of the *Mahonia* × *media* cultivars. The evergreen foliage is similar to other mahonias, with its holly-like leaves. Mulch around the plant in the spring.

3 NOVEMBER

Golden clematis *Clematis* 'Bill MacKenzie'

PLANT TYPE: Climber | **FAMILY:** Ranunculaceae | **HEIGHT:** 4.5m | **SPREAD:** 3m | **EXPOSURE:** Full sun, Partial shade | **ASPECT:** South, north, east or west facing

This vigorous clematis offers a long season of interest with flowers that begin in July and continue into November, followed by attractive puffball seedheads. It's also the perfect solution for covering a north-facing wall, where it will thrive where many other clematis would struggle as it can tolerate partial shade. The flowers are a sunny yellow, giving it the common name, golden clematis. It is low maintenance when it comes to pruning. Cut this clematis back hard in late February to a pair of low buds, about 30cm from the ground.

4 NOVEMBER

Cape primrose *Streptocarpus* 'Falling Stars'

PLANT TYPE: Perennial | **FAMILY:** Gesneriaceae | **HEIGHT:** 40cm | **SPREAD:** 45cm | **EXPOSURE:** Partial shade | **ASPECT:** South or east facing

Streptocarpus are tender perennials that make excellent house plants, because of their long flowering nature. This variety is incredibly long flowering, with light blue blooms from February to November. The flowers are small with a white centre, held above hairy leaves. Place in a spot that gets bright, indirect light, such as an east- or west-facing windowsill. In the winter, move to a brighter, warm place, such as a south-facing windowsill. Streptocarpus also benefit from humidity, so a kitchen or bathroom would be a good place to grow one, or the plant can be placed on a tray of moist pebbles.

---

5 NOVEMBER

Sowbread *Cyclamen cilicium*

PLANT TYPE: Perennial | **FAMILY:** Primulaceae | **HEIGHT:** 10cm | **SPREAD:** 10cm | **EXPOSURE:** Partial shade | **ASPECT:** East or west facing

The flowers of this cyclamen appear from September to November and are white to rose pink with a sweet honey scent and a darker magenta spot at the base of each petal. The petals have an attractive twisted habit and are paler than many autumn-flowering cyclamens. The oval or heart-shaped leaves provide interest through autumn and winter, with marbled patterns and a dark green blotch in the centre that looks like a Christmas tree. These dainty plants will grow well beneath deciduous trees and shrubs.

6 NOVEMBER

Regal pelargonium *Pelargonium* 'Lord Bute'

PLANT TYPE: Perennial | **FAMILY:** Geraniaceae | **HEIGHT:** 35cm | **SPREAD:** 40cm | **EXPOSURE:** Full sun | **ASPECT:** South or west facing

There are many forms of pelargoniums, with a variety of flower and leaf types. 'Lord Bute' is an old Regal pelargonium, which usually have large flowers borne in small clusters and scented leaves. 'Lord Bute' has purple-black flowers with pale purple edges from June to November, and fresh green foliage. The flowers can be used to decorate cakes or salads. Pelargoniums are tender and are often grown as annual bedding plants, but they can be overwintered indoors on a sunny windowsill or conservatory. Feed with a high-potash fertiliser during summer and deadhead regularly to keep plants looking neat.

7 NOVEMBER

African geranium *Pelargonium sidoides*

PLANT TYPE: Tender perennial | **FAMILY:** Geraniaceae | **HEIGHT:** 30cm | **SPREAD:** 45cm | **EXPOSURE:** Full sun | **ASPECT:** South, east or west facing

This is a species pelargonium that is native to South Africa, with beautiful, dark magenta flowers. It has long been used as a traditional medicine and is still used in cold and flu remedies in Europe due to its antibacterial effects. Its long-flowering nature makes it an attractive addition to container displays, where it will add colour from June to November. To keep plants healthy, water and feed them during the summer. Deadhead regularly to encourage more flowers. Before the frosts, lift plants if they are in the ground and pot them up to bring indoors. Cut the top growth back by a third.

---

8 NOVEMBER

Star grass *Dichromena colorata*

PLANT TYPE: Perennial | **FAMILY:** Cyperaceae | **HEIGHT:** 60cm | **SPREAD:** 45cm | **EXPOSURE:** Full sun, Partial shade | **ASPECT:** South, east or west facing

Dichromena colorata is a marginal pond plant that will also thrive in boggy areas of the garden. Marginal plants are best for shallow water or damp areas; star grass grows well in up to 5cm of water or boggy soil, as its natural home is the marshy coastal areas of North America. Its upright stems are topped by clusters of leaves. The white flowers are actually bracts (modified leaves), which appear from July to November and are star shaped, as the common name suggests. Tidy up the plant in autumn or spring once the new shoots appear, although spring is preferable as the old foliage will help protect any emerging shoots.

---

9 NOVEMBER

Daisy Bellissima Series *Bellis perennis* Belissima Series

PLANT TYPE: Perennial | **FAMILY:** Asteraceae | **HEIGHT:** 15cm | **SPREAD:** 15cm | **EXPOSURE:** Full sun, Partial shade | **ASPECT:** South, north, east or west facing

Bellis perennis Bellissima Series is a perennial, but it is often used as an annual and as bedding for its colourful and floriferous nature. These compact plants are useful for edging borders or pathways or for giving a boost to windowboxes and containers. This particular type flowers late, from September to November. Bellissima Series is a mix of daisies with pink, red and white double flowers. Not only does it flower in the autumn, it will overwinter and flower once more in the spring. Plant in August to September for autumn flowers.

---

10 NOVEMBER

Seven son flower tree *Heptacodium miconioides*

PLANT TYPE: Shrub | **FAMILY:** Caprifoliaceae | **HEIGHT:** 6m | **SPREAD:** 3m | **EXPOSURE:** Full sun, Partial shade | **ASPECT:** South, east or west facing

Trees with year-round interest in the garden are a valuable addition to the border – the seven son flower tree has many attributes, including scented flowers in late summer that last into November. Its glossy pairs of leaves are also an asset and after the flowers fall, there is colour from the pink-red calyxes that are left behind, which turn a red-purple shade. Through winter, the light bark is a feature, peeling away to reveal a dark shade underneath. *Heptacodium miconioides* can be grown as a shrub or a compact tree, making it ideal for small gardens.

11 NOVEMBER

Shrubby alcalthaea 'Parkallee' × *Alcalthaea suffrutescens* 'Parkallee'

PLANT TYPE: Perennial | **FAMILY:** Malvaceae | **HEIGHT:** 2.5m | **SPREAD:** 1.25m | **EXPOSURE:** Full sun | **ASPECT:** South or west facing

Shrubby alcalthaea 'Parkallee' is related to the hollyhock, but unlike hollyhocks it's reliably perennial and rust resistant. It's a prolific flowerer, with cup-shaped, apricot blooms appearing up its tall stems. The flowers appear between July and November and are an apricot colour. Cut the plants down to around 50cm from the ground or more if desired, once flowering is over in the autumn. Unlike hollyhocks, this plant has a branching habit, forming a vase shape, increasing in size with each year.

12 NOVEMBER

Sowbread *Cyclamen hederifolium* var. *hederifolium* f. *albiflorum*

PLANT TYPE: Bulb | **FAMILY:** Primulaceae | **HEIGHT:** 10cm | **SPREAD:** 10cm | **EXPOSURE:** Partial shade | **ASPECT:** South, north, east or west facing

Cyclamen hederifolium var. *hederifolium* f. *albiflorum* is also known as ivy-leaved cyclamen due to its marbled, ivy-like leaves. The upright white flowers appear before the leaves in October to November, a welcome sight in autumn in woodland or beneath deciduous trees and shrubs. They make attractive, useful groundcover, especially in dry areas beneath trees, and add interest to the autumn garden, alongside shade-loving plants such as evergreen ferns or in a carpet beneath trees such as acers with their fiery autumn leaf colour.

13 NOVEMBER

Viburnum *Viburnum × bodnantense* 'Charles Lamont'

PLANT TYPE: Deciduous shrub | **FAMILY:** Adoxaceae | **HEIGHT:** 2.5m | **SPREAD:** 1.5m | **EXPOSURE:** Full sun, Partial shade | **ASPECT:** South or west facing

As summer- and autumn-flowering perennials and shrubs come to the end of their season, *Viburnum × bodnantense* 'Charles Lamont' is just coming into flower. Its fragrant pink flowers are produced on bare branches between November and March, making it a spectacular feature plant on its own or as part of a mixed border. Ideally, place it near a front or back door, or a pathway that is used during winter, in order to enjoy its scent on a cold winter's day. Grow this shrub in moist, well-drained soil in a sunny sheltered site.

14 NOVEMBER

Autumn crocus *Crocus speciosus*

PLANT TYPE: Bulb | **FAMILY:** Iridaceae | **HEIGHT:** 15cm | **SPREAD:** 5cm | **EXPOSURE:** Full sun | **ASPECT:** South or west facing

One of the earliest autumn crocuses to flower, *Crocus speciosus* looks spectacular planted beneath deciduous trees and shrubs. From October to November, its blue to violet flowers create a colourful understorey in the border or in grass and will increase year on year. The goblet-shaped flowers are also pretty up close, with blue veins running through the violet petals, and they make a lovely addition to containers. In the ground, plant in groups, around 10cm deep. Crocus look best planted in drifts beneath trees and shrubs, rather than dotted around individually. If planted midsummer to early autumn, crocus should flower in the first year.

15 NOVEMBER

Saffron crocus *Crocus sativus*

PLANT TYPE: Bulb | **FAMILY:** Iridaceae | **HEIGHT:** 15cm | **SPREAD:** 5cm | **EXPOSURE:** Full sun | **ASPECT:** South facing

The stigma of *Crocus sativus* is used to make saffron, one of the most expensive spices you can buy. Although you could harvest your own, it would take around 150 flowers to create 1g of spice. Instead, enjoy the autumn colour that saffron crocuses bring to gardens, their rich purple flowers appearing at a time when many blooms are fading. It's a lovely bulbs for autumn container displays or for naturalising in grass, where they will spread to create a carpet of colour. Crocuses are a vital source of nectar and pollen for bumblebees. Plant the corms in August to September, in gritty, well-drained soil.

16 NOVEMBER

Belladonna lily *Amaryllis belladonna*

PLANT TYPE: Bulb | **FAMILY:** Amaryllidaceae | **HEIGHT:** 75cm | **SPREAD:** 30cm | **EXPOSURE:** Full sun | **ASPECT:** South or west facing

This outdoor amaryllis flowers October to November. Also known as the Jersey lily, it has funnel-shaped pink blooms at the top of bare purple stems, as the flowers appear before the leaves. The flowers are large, around 10cm long, and scented, making a showstopping display in autumn. They go well with silver-leaved plants such as caryopteris. Although this amaryllis is hardy through most of the UK, in cold regions it should be planted in the shelter of a south-facing wall or in a warm sunny spot. Don't plant the bulbs too deeply, just below the soil surface. It can also be grown in pots and brought indoors over winter.

17 NOVEMBER

Clematis *Clematis* 'Madame Julia Correvon'

PLANT TYPE: Deciduous climber | **FAMILY:** Ranunculaceae | **HEIGHT:** 3m | **SPREAD:** 1.5m | **EXPOSURE:** Full sun, Partial shade | **ASPECT:** South, north, east or west facing

Clematis 'Madame Julia Correvon' was raised in France in 1900, but was lost to cultivation until it was rediscovered by plantsman Christopher Lloyd. This late clematis flowers from July until November, covering boundaries in large, claret-red, star-shaped flowers. It will grow up to 3m tall, perfect for softening fences and walls with its bright blooms. Viticella clematis are easy to look after, as the pruning is easy – simply cut back to a pair of buds, around 30cm from the ground in February or March. Plant it in the spring and it should flower by late July.

---

18 NOVEMBER

Peruvian lily *Alstroemeria* 'Avrille' Majestic Series

PLANT TYPE: Perennial | **FAMILY:** Alstromeriaceae | **HEIGHT:** 90cm | **SPREAD:** 30cm | **EXPOSURE:** Full sun | **ASPECT:** South, east or west facing

Alstroemeria originates from South America and is also known as Lily of the Incas. It has showy and brightly coloured flowers that can include red, orange, yellow, purple as well as lighter colours like pink and white. They have an exotic look, making them popular both as a border flower and a cut flower, as the blooms are long lasting. Many varieties, such as 'Maxi Avrille' have been bred to be long flowering and will last into November. This variety has peach-pink flowers with yellow throats. Plant in a sheltered sunny spot or in containers.

---

19 NOVEMBER

Penstemon *Penstemon* 'Raven'

PLANT TYPE: Perennial | **FAMILY:** Plantaginaceae | **HEIGHT:** 1m | **SPREAD:** 30cm | **EXPOSURE:** Full sun, Partial shade | **ASPECT:** South, east or west facing

Penstemons are incredibly long flowering, beginning in July and often carrying on until the first frosts. They are easy to grow and a good pick for the middle of a mixed border or a container display, with foxglove-like flowers. These are attractive to bumblebees and valuable due to the long flowering period. 'Raven' has deep purple flowers making a good plant partner for purple-flowering perennials as well as foliage plants such as euphorbia, which have lime green leaves and flowers. To keep them flowering into November, deadhead your penstemons regularly. Divide clumps in the spring if they are congested.

---

20 NOVEMBER

Candy cane sorrel *Oxalis versicolor*

PLANT TYPE: Bulb | **FAMILY:** Oxalidaceae | **HEIGHT:** 8cm | **SPREAD:** 20cm | **EXPOSURE:** Full sun, Partial shade | **ASPECT:** South, east or west facing

Oxalis versicolor has white flowers with red stripes that curl around. Although only 8cm tall, these bulbs provide immediate impact, with their striking colouring. The flowers open up during the day and close up at night. Oxalis bulbs will spread easily, creating attractive groundcover. To restrict them, plant the bulbs in a container – this also has the benefit of being able to move indoors, as oxalis is frost tender. For any bulbs growing in a border, cover with a layer of mulch in the autumn to protect from the cold.

21 NOVEMBER

Himalayan fairy grass *Miscanthus nepalensis*

PLANT TYPE: Perennial grass | **FAMILY:** Poaceae | **HEIGHT:** 1.5m | **SPREAD:** 1.2m | **EXPOSURE:** Full sun | **ASPECT:** South, north, east or west facing

Grasses add movement to borders, and in the autumn their flowers and seedheads complement the colours of the season. This miscanthus has graceful golden flower plumes from late summer, while the seedheads last into winter, adding architectural interest. This ornamental grass would go well with late-flowering perennials such as asters or rudbeckias, its golden flowers and seedheads complementing the rich colour of neighbouring flowers. Grow in a sheltered spot and cut back to ground level in the spring to make way for new growth.

22 NOVEMBER

Masson's wood sorrel *Oxalis massoniana*

PLANT TYPE: Bulb | **FAMILY:** Oxalidaceae | **HEIGHT:** 10cm | **SPREAD:** 10cm | **EXPOSURE:** Full sun | **ASPECT:** South, east or west facing

This dainty alpine looks good planted in shallow pots that can be displayed on outdoor tables or greenhouse shelving, where the flowers can be admired up close. The flower buds are a dark orange, opening up into pale orange flowers with a yellow centre. *Oxalis massoniana* is from South Africa and can cope with low temperatures but won't survive frost. In cold regions it can be grown in a cold greenhouse or alpine house, or covered outdoors when the frost is forecast. Plant the bulbs 5cm deep in potting compost with added grit or in a rockery in areas where the temperature doesn't fall below zero.

23 NOVEMBER

Anchor plant *Colletia paradoxa*

PLANT TYPE: Deciduous shrub | **FAMILY:** Rhamnaceae | **HEIGHT:** 3m | **SPREAD:** 5m | **EXPOSURE:** Full sun | **ASPECT:** South or west facing

The flowers of *Colletia paradoxa* are produced on a dangerously spiny bush and have a sweet, almond-like scent. This is a very unusual shrub, which will make a talking point in the garden. The branches, which look like leaves, are triangular with spines on the end, so it would make a good plant to keep burglars out and prevents any animals feeding on the leaves. The tiny bell-shaped flowers appear from autumn into winter. This plant is native to Brazil and Uruguay and was introduced into Britain in around the 1820s.

24 NOVEMBER

Marguerite *Argyranthemum* 'Cornish Gold'

PLANT TYPE: Perennial | **FAMILY:** Asteraceae | **HEIGHT:** 70cm | **SPREAD:** 70cm | **EXPOSURE:** Full sun | **ASPECT:** South or east facing

Compact and incredibly long flowering, *Argyranthemum* 'Cornish Gold' produces a mass of yellow daisies from May until November. These will attract a wide range of insects. The golden-yellow flowers are 5cm across on a plant that's only 60cm tall, a perfect plant for a container or for filling gaps at the front of the border. There are many colours of marguerite to choose from including pink, white and yellow. Marguerite is tender, so will need to be moved indoors to overwinter. Alternatively, take summer cuttings to grow new plants for free.

25 NOVEMBER

Korean clematis *Clematis serratifolia*

PLANT TYPE: Climber | **FAMILY:** Ranunculaceae | **HEIGHT:** 4m | **SPREAD:** 1.5m | **EXPOSURE:** Full sun, Partial shade | **ASPECT:** South or east facing

This Korean clematis was brought back from Korea by the famous plant hunter EH Wilson in around 1918. Its flowers are a pale, lemon yellow and appear over a long period from July to November. Even after this, the plant continues to add interest with silky seedheads that last well into winter. When planted in a sunny position, the citrus scent of its flowers are stronger, and it would be a lovely choice for a climber on a fence near a seating area. Keep the bottom of the plant in shade with other well-placed plants or by putting flat stones over the soil where it is planted.

26 NOVEMBER

Backhouse Australian fuchsia *Correa backhouseana*

PLANT TYPE: Evergreen shrub | **FAMILY:** Rutaceae | **HEIGHT:** 2m | **SPREAD:** 1.5m | **EXPOSURE:** Full sun | **ASPECT:** South or east facing

Correa backhouseana was discovered by botanist Sir Joseph Banks on an expedition to Australia's Botany Bay aboard the HMS *Endeavour*. In Australia it's a hardy plant, but in Britain it was a popular conservatory plant during the 18th century, producing bell-like yellow to green flowers from late autumn through to spring. There are many species of Correa including varieties with a variety of flower shades including red, pink, yellow and green flowers. *Correa backhouseana* is hardy to about −5°C, which means it can be grown outdoors, but in cold regions it may need to be grown indoors.

27 NOVEMBER

Aniseed boronia *Boronia crenulata*

PLANT TYPE: Shrub | **FAMILY:** Rutaceae | **HEIGHT:** 90cm | **SPREAD:** 90cm | **EXPOSURE:** Full sun, Partial shade | **ASPECT:** South, east or west facing

The scent of *Boronia crenulata* comes from its aromatic foliage – a fragrant mix of aniseed and fennel. Its pink flowers are in bloom from July to December, covering the compact shrub. This plant is native to Western Australia and used to growing in warm conditions, so it is suited to a mild location or a cool greenhouse as it can only tolerate temperatures down to 1°C. Plant in ericaceous compost and shade from hot sun, if grown indoors. It would be a good plant for coastal areas or on a patio during frost-free months.

28 NOVEMBER

Long leaf wax flower *Eriostemon myoporoides*

PLANT TYPE: Evergreen shrub | **FAMILY:** Rutaceae | **HEIGHT:** 2m | **SPREAD:** 2m | **EXPOSURE:** Full sun | **ASPECT:** South, east or west facing

Also known as the gin and tonic plant, because of its fragrance that is apparently reminiscent of a gin and tonic. The attraction of this plant is that it not only flowers in winter, but will bloom again in the spring, with tiny, white star-shaped flowers covering the compact shrub. This is a popular plant with florists because of the long-lasting flowers and scent. The flowers are also attractive to pollinators. It's not fully hardy so it also makes a good conservatory plant, but it can be grown outdoors in a sheltered position, in mild regions of the UK where it will not be affected by frost.

29 NOVEMBER

Rosemary grevillea *Grevillea rosmarinifolia*

PLANT TYPE: Evergreen shrub | **FAMILY:** Proteaceae | **HEIGHT:** 2.5m | **SPREAD:** 4m | **EXPOSURE:** Full sun | **ASPECT:** South, east or west facing

Rosemary grevillea looks similar to rosemary with its evergreen, needle-like leaves, but it's more vigorous and flowers in winter. These flowers are a deep red, held at the end of its branches in clusters, a welcome sight on a grey November day. In sheltered areas it may also flower in spring and summer. This shrub thrives in poor, sandy or peaty soil and will not grow well on rich, fertile ground. It can spread up to 4m, and can also be grown as a hedge. Prune to tidy the shrub in early spring.

---

30 NOVEMBER

Chilean bellflower *Lapageria rosea*

PLANT TYPE: Evergreen climber | **FAMILY:** Philesiaceae | **HEIGHT:** 4m | **SPREAD:** 2.5m | **EXPOSURE:** Full shade, Partial shade | **ASPECT:** North, east or west facing

Lapageria rosea was introduced by plant hunter William Lobb in 1840. As indicated by its common name, this colourful climber is native to Chile, and it will only thrive outdoors in areas where there is no frost, otherwise it should be grown in a conservatory or cool greenhouse. In mild areas, it is a colourful focal point for shady areas into late autumn. Through July to November, it has long pink flowers on twining evergreen stems that can grow up to 4m long. It can be pruned after flowering to keep the climber compact, but this is not essential.

DECEMBER

1 DECEMBER

Viburnum *Viburnum tinus*

PLANT TYPE: Evergreen shrub | **FAMILY:** Viburnaceae | **HEIGHT:** 3m | **SPREAD:** 3m | **EXPOSURE:** Full shade, Partial shade, Full sun | **ASPECT:** South, east or west facing

A long-flowering winter shrub is a wonderful addition to the border, blooming at a time when few other shrubs are in flower. Viburnums can grow into large shrubs, providing shelter for wildlife as well as making an attractive focal point. The flowerheads are flat, with clusters of small, creamy white flowers that are good for early bees foraging for nectar. This shrub can also be grown as a hedge. A popular cultivar is 'Eve Price', which forms a more compact shrub than the species. Grow in moist, well-drained soil and mulch annually with well-rotted manure or compost. Reshape the plant in early summer.

2 DECEMBER

Japanese mahonia *Mahonia japonica*

PLANT TYPE: Evergreen shrub | **FAMILY:** Berberidaceae | **HEIGHT:** 2m | **SPREAD:** 3m | **EXPOSURE:** Full shade, Partial shade, Full sun | **ASPECT:** South, north, east or west facing

This mahonia species is native to China, not Japan, as its common name suggests, but it has been cultivated in Japan for centuries. Its flowers have a strong fragrance and vibrant yellow colour, giving it stand-out value in a winter border. For the gardener it has the added bonus of being low-maintenance and evergreen, bringing interest all year round with its long glossy leaves – these can be up to 45cm long – and thriving in shady problem spots. The flowers too are held on long racemes of up to 25cm, over a long period from November to March.

3 DECEMBER

Perennial wallflower *Erysimum* 'Bowles's Mauve'

PLANT TYPE: Evergreen perennial | **FAMILY:** Brassicaceae | **HEIGHT:** 75cm | **SPREAD:** 60cm | **EXPOSURE:** Full sun | **ASPECT:** South, east or west facing

The perennial wallflower *Erysimum* 'Bowles's Mauve' should be in everyone's garden. It has spires of vivid purple flowers that last for months. In southern parts of the UK, it can flower into winter. Perhaps because of this highly floriferous nature, *Erysimum* 'Bowles's Mauve' is short-lived, but it is easy to create new plants from cuttings. As well as its attractive flowers, the foliage is evergreen, with narrow blue-green leaves. Deadhead flower stalks regularly to keep the plant blooming. Grow in the border or use as a feature plant in a container.

4 DECEMBER

Sweet box *Sarcococca confusa*

PLANT TYPE: Evergreen shrub | **FAMILY:** Buxaceae | **HEIGHT:** 2m | **SPREAD:** 1m | **EXPOSURE:** Full shade, Partial shade | **ASPECT:** North, east or west facing

Tricky shady spots can be enhanced by this winter-flowering shrub that has sweetly scented, pure white flowers. As well as deep shade, it can tolerate pollution, dry shade and it's low maintenance as it can cope with neglect on the watering front. It is a good choice for woodland areas or shady borders, in particular near seating areas or paths where it's possible to enjoy the fragrance. Although it flowers from December to March, it's also evergreen, adding structure to borders throughout the year. It makes a good alternative to box, and can also be grown as a low hedge. Grow in well-drained soil.

---

5 DECEMBER

Christmas cactus *Schlumbergera*

PLANT TYPE: Perennial house plant | **FAMILY:** Cactaceae | **HEIGHT:** 40cm | **SPREAD:** 40cm | **EXPOSURE:** Partial shade | **ASPECT:** South, north, east or west facing

Schlumbergera flower over the Christmas period, from late November to January, hence the common name. Their trumpet-shaped flowers come in shades from red and purple through to white, a good plant for brightening up rooms indoors. In the wild, *Schlumbergera* grow in tropical rainforests, attached to trees. This means in our homes they do best in humid conditions, and of direct sunlight. If a humid spot isn't available, mist the plant regularly. To keep the plant flowering each year, it needs two rest periods with lower temperatures and less watering – after flowering in late winter and in September.

6 DECEMBER

Indoor cyclamen *Cyclamen persicum* varieties

PLANT TYPE: Bulb | **FAMILY:** Primulaceae | **HEIGHT:** 10cm | **SPREAD:** 15cm | **EXPOSURE:** Full sun, Partial shade | **ASPECT:** East or west facing

Indoor cyclamen can have flowers in many colours, including pink, red, violet and white and can flower in autumn, spring or winter depending on which varieties you have. The flowers can be frilly or scented, held on slim stems above attractive heart-shaped leaves. They should flower for about six weeks and die back in the spring ready for a period of dormancy over the summer. It is not dead and will flower again the following winter. Place in a spot that gets bright indirect light, but out of direct sunlight. Let the top few centimetres of compost dry out before watering again.

7 DECEMBER

Poinsettia *Euphorbia pulcherrima*

PLANT TYPE: Houseplant | **FAMILY:** Euphorbiaceae | **HEIGHT:** 60cm | **SPREAD:** 30cm | **EXPOSURE:** Full sun, Partial shade | **ASPECT:** South or west facing

The colour of poinsettias comes from the red bracts, which look like petals – the actual flowers are small and yellow, at the centre of these bracts. *Euphorbia pulcherrima* is native to Mexico, where it grows up to 2m tall, but as a house plant it's compact. Grow in bright light, away from direct sunlight. Water sparingly and mist regularly. Poinsettias are often thrown away after Christmas, but they can be reddened for the next year by pruning hard in April, to 10cm, repotted, then kept in a cool, north-facing windowsill over summer. From November, place in a dark room for 12 hours a day to encourage flowering.

8 DECEMBER

Paperwhite daffodil *Narcissus papyraceus* 'Ziva'

PLANT TYPE: Bulb | **FAMILY:** Amaryllidaceae | **HEIGHT:** 40cm | **SPREAD:** 8cm | **EXPOSURE:** Full sun, Partial shade | **ASPECT:** South, east or west facing

Bulbs that can be forced for the festive period are great for introducing colour and scent to your home in time for Christmas. Paperwhite narcissus has the strongest fragrance of all the narcissi and is traditionally used as a forced bulb to create table displays for the festive period or as cut flowers. Each stem has up to 10 white flowers. *Narcissus* 'Ziva' is a multi-headed, white daffodil that makes a wonderful house plant during the winter, flowering between December and January. The bulbs can be planted in the autumn, four to six weeks before you want them to flower.

9 DECEMBER

Forced hyacinth *Hyacinthus*

PLANT TYPE: Bulb | **FAMILY:** Asparagaceae | **HEIGHT:** 30cm | **SPREAD:** 15cm | **EXPOSURE:** Full sun, Partial shade | **ASPECT:** South, north, east or west facing

Hyacinths can also be forced to flower in time for Christmas, with each variety taking slightly different times to reach flowering. In general, they need to be planted around the end of September in order to flower at Christmas. They come in a range of colours, from the popular pink and purple types to white and even mauve. Hyacinth bulbs can be grown in baskets or individual vases of water, making attractive table displays or gifts. They need to be kept somewhere dark and cool until shoots appear, then moved into a light spot to encourage flowering.

10 DECEMBER

Indian azalea *Rhododendron simsii* **varieties**

PLANT TYPE: Shrub | **FAMILY:** Ericaceae | **HEIGHT:** 60cm | **SPREAD:** 60cm | **EXPOSURE:** Partial shade | **ASPECT:** North or west facing

This is a tender, indoor azalea that can not be grown outdoors in areas that are prone to frost. These plants are usually forced into flowering in time for Christmas. It has flowers in shades of pink, white and red and makes a good pot plant through winter. They can be moved outdoors during the summer and brought back indoors again in the autumn. It grows best in bright, indirect light in a cool, humid room. Let the top few centimetres of compost dry out before watering again. Water less frequently during winter.

11 DECEMBER

Butterfly amaryllis *Hippeastrum papilio*

PLANT TYPE: Bulb | **FAMILY:** Amaryllidaceae | **HEIGHT:** 50cm | **SPREAD:** 20cm | **EXPOSURE:** Full sun, Partial shade | **ASPECT:** South, east or west facing

This amaryllis is a bit different to your usual Christmas amaryllis, in that *papilio* has a few shorter stems instead of one tall stem. These are topped by showstopping white and red, trumpet-shaped flowers from December to January. The name 'butterfly amaryllis' comes from the shape of the lower petals, which look like a butterfly's wings. Plant the bulb so that the top two-thirds of the bulb is poking out of the compost. Keep the pot in bright, indirect light and turn it every so often to prevent the stems growing in one direction.

12 DECEMBER

Flamingo flower *Anthurium andreanum*

PLANT TYPE: House plant | **FAMILY:** Araceae | **HEIGHT:** 50cm | **SPREAD:** 40cm | **EXPOSURE:** Full shade, Partial shade | **ASPECT:** South, east or west facing

The flamingo flower has arrow- or heart-shaped leaves that are glossy and evergreen, providing interest indoors year round. It also provides exotic colour in pink, white or red, although the waxy, flower-like leaves are actually bracts and the real flower is the yellow 'tail' within. It flowers on and off throughout the year, which is a large part of its appeal for houseplant lovers. Flamingo flowers are native to tropical regions of South America, and thrive in warm, humid rooms, so a bathroom or steamy kitchen would be the ideal location. Keep out of direct sunlight.

13 DECEMBER

Winter honeysuckle *Lonicera × purpusii* 'Winter Beauty'

PLANT TYPE: Shrub | **FAMILY:** Caprifoliaceae | **HEIGHT:** 2m | **SPREAD:** 2.5m | **EXPOSURE:** Full sun, Partial shade | **ASPECT:** South, north, east or west facing

Lonicera × purpusii 'Winter Beauty' has a sweet fragrance from its winter flowers that appear on bare stems from December until March. The flowers perk up a border during cold months and provide valuable nectar for any bumblebees that have been disturbed during hibernation. The best place for this plant is near a path or entrance where its scent can be appreciated. As it only has one season of interest, it's a good choice for a large garden, or a mixed border where other plants can take over the show in spring or summer. This shrub flowers best in full sun.

14 DECEMBER

Winter-flowering clematis *Clematis napaulensis*

PLANT TYPE: Climber | **FAMILY:** Ranunculaceae | **HEIGHT:** 3m | **SPREAD:** 1m | **EXPOSURE:** Full sun, Partial shade | **ASPECT:** South, east or west facing

Winter-flowering clematis such as *Clematis napaulensis* are a gift in the cooler months, for boosting colour on boundaries or using as climbers in conservatories. This plant orgianated from Nepal and is classed as evergreen, but it does lose its leaves during the spring and summer when it goes dormant. Despite a lack of foliage during spring, this plant is well worth buying for the winter flowers alone; they are bell-shaped with striking red-purple stamens that stand out all the more against the pale greeny yellow petals. These pendulous flowers will make a statement in any sheltered garden or conservatory.

---

15 DECEMBER

Moth orchids *Phalaenopsis* varieties

PLANT TYPE: Houseplant | **FAMILY:** Orchidaceae | **HEIGHT:** 50cm | **SPREAD:** 40cm | **EXPOSURE:** Partial shade | **ASPECT:** North, east or west facing

Phalaenopsis, known as moth orchids, are easy to grow, with many different types and sizes available, with flower colours ranging from white to pink, through to those with spotted patterns. They make excellent house plants, flowering up to three times a year, with long-lasting, graceful flower heads. Moth orchids need a minimum temperature of 16°C in bright, indirect light – an east- or north-facing windowsill is ideal. They also thrive in humidity, so a kitchen or bathroom would be perfect. Water once a week with rainwater.

16 DECEMBER

Water hawthorn *Aponogeton distachyos*

PLANT TYPE: Aquatic plant | **FAMILY:** Aponogetonaceae | **HEIGHT:** 10cm | **SPREAD:** 90cm | **EXPOSURE:** Full sun, Partial shade | **ASPECT:** South, east or west facing

Water hawthorn is a deep-water plant for the surface of ponds, attracting wildlife such as dragonflies and butterflies with its scented flowers. It flowers twice a year, once from March to May and again in the autumn, carrying on until the first frosts. During the summer it is dormant and its foliage vanishes from the pond surface. Add in some feed when it starts to regrow and it should produce more flowers. This is a useful plant to increase surface coverage of your pond.

17 DECEMBER

Winter-flowering clematis *Clematis* 'Christmas Surprise'

PLANT TYPE: Climber | **FAMILY:** Ranunculaceae | **HEIGHT:** 2.5m | **SPREAD:** 2m | **EXPOSURE:** Full sun, Partial shade | **ASPECT:** South, east or west facing

This is a fairly new variety of winter-flowering clematis that blooms in time for Christmas and keeps going until February. The plant produces a mass of nodding bell-shaped flowers, which are a creamy white, with a tiny bit of pink in the centre. It's a good pick for a south-facing wall or fence, but it can also be grown up trees or shrubs to add interest through the colder months. This clematis is frost hardy, so it might need protection in cold weather. It's low maintenance as it doesn't need annual pruning, just a light tidy if the growth needs to be restricted.

---

18 DECEMBER

Winter-flowering cherry *Prunus × subhirtella* 'Autumnalis Rosea'

PLANT TYPE: Deciduous tree | **FAMILY:** Rosaceae | **HEIGHT:** 4m | **SPREAD:** 4m | **EXPOSURE:** Full sun | **ASPECT:** South, north, east or west facing

In mild weather, this cherry produces rose pink, double flowers between November and March. The leaves turn orange and drop just as the tree is beginning to flower. It will then flower on and off through to early spring. *Prunus × subhirtella* is a species native to Japan – it was introduced to Britain at the beginning of the 20th century. It also has a white-flowering version, *Prunus × subhirtella* 'Autumnalis'. The flowers are followed by small fruits that the birds enjoy. This is the perfect tree for a small garden, offering several seasons of interest.

---

19 DECEMBER

Arrowwood 'Dawn' *Viburnum × bodnantense* 'Dawn'

PLANT TYPE: Deciduous shrub | **FAMILY:** Viburnaceae | **HEIGHT:** 3m | **SPREAD:** 2m | **EXPOSURE:** Full sun, Partial shade | **ASPECT:** South, north, east or west facing

Viburnum 'Dawn' is a wonderful winter-flowering shrub, carrying clusters of pink flowers on naked stems. Along with the thrill of flowers in winter, this shrub brings a sweet scent to the garden. The flowering period is long, lasting from November through to early spring, when the attractive, toothed green leaves begin to unfurl and the flowers fade to white. For the rest of the year, this shrub provides structure and height in borders. In autumn the leaves turn orange and yellow before falling, giving it another season of interest.

---

20 DECEMBER

Winter-flowering clematis *Clematis cirrhosa* var. *purpurascens* 'Freckles'

PLANT TYPE: Evergreen climber | **FAMILY:** Ranunculaceae | **HEIGHT:** 4m | **SPREAD:** 1.5m | **EXPOSURE:** Full sun | **ASPECT:** South or west facing

'Freckles' is a very popular winter-flowering climber, due to its speckled red and cream flowers that appear between the months of October and February. To admire these flowers close up, plant over an arch or somewhere that the inside of these nodding bell-shaped flowers can be admired. Winter clematis are a great source of nectar for winter-active bumblebees as well as covering boundaries during a quiet time of year on the flowering front. The bonus of this clematis is that it's evergreen, creating year-round interest, also useful if you have trellis as it will provide extra privacy.

21 DECEMBER

Camellia *Camellia × vernalis* 'Yuletide'

PLANT TYPE: Evergreen shrub | **FAMILY:** Theaceae | **HEIGHT:** 2.5m | **SPREAD:** 1.5m | **EXPOSURE:** Full shade, Partial shade | **ASPECT:** North or west facing

Camellias are native to South Asia where they grow in woodland, so they do best in shade, making them useful plants for adding colour in gloomy spots. 'Yuletide' is covered with bright red flowers from November through to February, peaking around Christmastime – hence the common name. Camellias, if planted in a suitable location, are low-maintenance shrubs that bring year-round interest to the garden with their evergreen foliage. Plant in acid soil in a sheltered, shady spot. Camellia are usually best in west- or north-facing positions, as morning sun can damage the flowerbuds.

22 DECEMBER

Amaryllis *Hippeastrum* (Galaxy Group) 'Spartacus'

PLANT TYPE: Bulb | **FAMILY:** Amaryllidaceae | **HEIGHT:** 50cm | **SPREAD:** 20cm | **EXPOSURE:** Bright filtered light | **ASPECT:** East or west facing

Add vibrant colour to your home in December with a spectacular amaryllis – its flowers are festive-coloured, with red and white flower heads. Each flower stem has around four flowers and they can grow to 15–20cm in diameter. The red petals have a pattern in the shape of a star. It's easy to grow amaryllis – simply plant up a bulb eight weeks before Christmas with the top of the bulb sitting just above the rim of the pot. This variety will flower from December through to February, depending on when the bulb is planted. They generally flower six to eight weeks after planting.

23 DECEMBER

Aniseed buchu *Agathosma capensis*

PLANT TYPE: Evergreen shrub | **FAMILY:** Rutaceae | **HEIGHT:** 30cm | **SPREAD:** 30cm | **EXPOSURE:** Full sun | **ASPECT:** South, east or west facing

Agathosma capensis originates in South Africa, and is a source of Buchu, which is an antiseptic essential oil. It's a compact evergreen shrub that has needle-like, fragrant leaves and small flowers that vary in shade, from mauve through to pink and white – each tiny flower has five petals. This shrub can't tolerate temperatures below 5°C so should be grown in a conservatory or cool greenhouse, where it makes an attractive feature, or alternatively be brought indoors over winter.

---

24 DECEMBER

Mount Noko camellia *Camellia transnokoensis*

PLANT TYPE: Evergreen shrub | **FAMILY:** Theaceae | **HEIGHT:** 3m | **SPREAD:** 3m | **EXPOSURE:** Partial shade | **ASPECT:** North or west facing

The flowers on this upright camellia are small and single, with a sweet scent. They open from buds that have a pink tip, usually starting in December and continuing through into early spring, sometimes into March. *Camellia transnokoensis* comes from the island of Taiwan, and has a graceful growing habit, making it a striking feature plant. The leaves are small, emerging bronze but glossy for the rest of the year, adding interest year round. Once established this camellia will survive harsh winters in the south but may need protection in northern counties.

---

25 DECEMBER

Cone bush *Isopogon formosus*

PLANT TYPE: Evergreen shrub | **FAMILY:** Proteaceae | **HEIGHT:** 1.5m | **SPREAD:** 1m | **EXPOSURE:** Full sun | **ASPECT:** South, east or west facing

Isopogon formosus gets its common name from the cones that form after the flowers. There's no missing the showy, bright pink flowers, bound to brighten any December day, which look like sea anemones when they first emerge. Native to Australia, this shrub does best in mild and coastal areas of the UK with shelter from cold winds. It may not survive harsh winters where temperatures drop further than −5°C. For areas that colder than this, it is better to grow it in an unheated greenhouse.

26 DECEMBER

Hellebore *Helleborus × ballardiae* HGC Snow Dance

PLANT TYPE: Perennial | **FAMILY:** Ranunculaceae | **HEIGHT:** 40cm | **SPREAD:** 50cm | **EXPOSURE:** Full sun, Partial shade | **ASPECT:** South, east or west facing

Snow Dance is an early flowering hellebore variety, with large white flowers that face outward. This is unusual in a hellebore, which often have downward-facing flowers – an attractive attribute that enables you to admire the blooms easily. They are good container plants, ideal for adding colour to winter displays on the doorstep or in a windowbox. Alternatively, they could be planted beneath deciduous trees or shrubs to create winter interest. They are loved by pollinators and have foliage that is evergreen, providing groundcover for much of the year.

27 DECEMBER

Polyspora speciosa

PLANT TYPE: Shrub or small tree | **FAMILY:** Theaceae | **HEIGHT:** 2m | **SPREAD:** 1m | **EXPOSURE:** Full sun, Partial shade | **ASPECT:** South, east or west facing

The flowers of *Polyspora speciosa* looks similar to a camellia and has long glossy leaves up to 20cm long. The white flowers are large, too, up to 15cm across. Although not as well known as camellias, it is just as attractive, hardy down to −10°C and looks good year round due to its attractive foliage. The flowers appear from December through to March, a bonus to a winter garden. This would make a wonderful feature plant. They have peeling bark and the flowers are followed by fruit that looks similar to an acorn. Plant it in a warm spot in neutral to acid soil.

28 DECEMBER

Iris 'Mary Barnard' *Iris unguicularis* 'Mary Barnard'

PLANT TYPE: Bulb | **FAMILY:** Iridaceae | **HEIGHT:** 40cm | **SPREAD:** 30cm | **EXPOSURE:** Full sun | **ASPECT:** South facing

The rich violet flowers of Iris 'Mary Barnard' not only provide a blast of colour in December but a strong fragrance, too. Cut the flowers to bring indoors where the perfume can be appreciated. This iris is native to Algeria and Tunisia and grows well in full sun, particularly thriving in the shelter of a south-facing wall. If the soil is a heavy clay, it may be better to plant the bulbs in well-draining compost with added grit. It will flower prolifically all winter. Tidy the foliage after flowering.

---

29 DECEMBER

Glory bush *Tibouchina paratropica*

PLANT TYPE: Perennial | **FAMILY:** Melastomataceae | **HEIGHT:** 60cm | **SPREAD:** 60cm | **EXPOSURE:** Full sun | **ASPECT:** South facing

This is an exotic perennial from South America that is borderline hardy, which means it can be grown outdoors in mild areas – in the shelter of a south-facing wall or somewhere similarly sheltered, but in colder regions needs to be grown in a conservatory or greenhouse. It has clusters of flowers that emerge in a variety of shades of pink. It begins flowering in late summer and goes on into December. Typically it's described as growing to around 60cm, although in the right location it could grow bigger than this.

30 DECEMBER

Cootamundra wattle *Acacia baileyana* 'Purpurea'

PLANT TYPE: Evergreen shrub | **FAMILY:** Fabaceae | **HEIGHT:** 5m | **SPREAD:** 3m | **EXPOSURE:** Full sun, Partial shade | **ASPECT:** South or west facing

Acacia baileyana 'Purpurea' can be a small tree or large shrub, it comes from Australia and grows well in warm or coastal gardens, flowering from December to April. It's also known as golden mimosa because of its mass of yellow fluffy flowers that have a sweet scent and attract early pollinators. The foliage is just as attractive as the flowers, fern-like blue to green leaves with a purple hue. It can be grown outdoors in mild areas in a sheltered spot, or indoors in a greenhouse or as a house plant.

31 DECEMBER

Camellia 'Sparkling Burgundy' *Camellia sasanqua* 'Sparkling Burgundy'

PLANT TYPE: Evergreen shrub | **FAMILY:** Theaceae | **HEIGHT:** 3m | **SPREAD:** 2m |
EXPOSURE: Full shade, Partial shade | **ASPECT:** North or west facing

Camellias offer many attributes to the garden, including evergreen foliage and structure, but their main appeal is their flowers. This autumn-flowering camellia comes from North America and its scented, double flowers are stunning. They are a rich pink and peony shaped, produced generously over autumn and winter. Once established this camellia can also tolerate a sunny spot, as long as its roots are kept cool. Plant it in acid to neutral soil.

INDEX

Note: page numbers in **bold** refer to illustrations.

Abelia x *grandiflora* 186, **186**
Abeliophyllum distichum 17, **17**
Abutilon megapotamicum 198, **198**
Abyssinian gladiolus 192
Acacia
 A. baileyana `Purpurea' 248, **248**
 A. dealbata (mimosa) 16, **16**
Acanthus
 A. mollis 141, **141**
 A. spinosus 99, **99**
Achillea 98, 163, 172, 186
 A. millefolium 117, **117**
acidic soil 14, 16, 18, 26, 37, 48, 69, 77, 84–5, 156, 169, 243, 246, 249
 see also ericaceous plants
aconitum 174
African geranium 215, **215**
Agastache `Blue Fortune' 192, **192**
Agathosma capensis 245, **245**
Agrostemma githago 147, **147**
ajuga 25
Akebia quinata 83, **83**
Alcalthaea suffrutescens `Parkallee' 217, **217**
Alcea `Sunshine' 136, **136**
Algerian iris 43, **43**
alkaline soils 14, 48
Allium 116, 146
 A. cristophii 132, **132**
 A. stipitatum 92, **92**
Alstroemeria
 A. `Avrille' Majestic series 221, **221**
 A. `Cahors' 166, **166**
amaranthus 133
Amarine `Antastsia' (Belladiva Series) 198, **198**
amaryllis (*Hippeastrum*) 28, **28**, 29, **29**, 31, 244, **244**
 butterfly amaryllis 237, **237**
Amaryllis belladonna (belladonna lily) 198, 220, **220**
Amira maritima 98, **98**
Ammi majus 139, **139**
anchor plant 225, **225**
Anderson's hebe 160, **160**
Anemone
 A. nemorosa 55, **55**
  A. n. `Blue Eyes' 72, **72**
 A. x *hybrida*
  A. x *h.* `Honorine Jobert' 161, **161**
  A. x *h.* `September Charm' 180, **180**
Angelica archangelica 126, **126**
anise scented sage 182, **182**
aniseed boronia 227, **227**
aniseed buchu 245, **245**
annuals 15, 17, 28, 101
 August-flowering 158, 168
 hardy 87, 97, 101, 119, 120
 July-flowering 133, 138, 139, 141–3, 147–8
 June-flowering 119–20, 125
 May-flowering 97, 101, 105, 106
 October-flowering 197, 199, 205
 September-flowering 173, 178
Annunciation lily *see* Madonna lily
Anthophora plumipes 73
Anthurium andreanum 238, **238**
Antirrhinum majus `Royal Bride' 106, **106**
Aponogeton distachos 240, **240**
April 70–89
aquatic plants 153, 240
Aquilegia flabellata 94, **94**
Arbutus unedo 201, **201**
Argentinian vervain (*Verbena bonariensis*) 135, **135**, 152, 172, 175, **175**
Argyranthemum frutescens `Cornish Gold' 226, **226**
Armand clematis 81, **81**
arrowwood `Dawn' 241, **241**
asters 154, 175, **175**, 186, 195, **195**
Astrantia major `Hadspen Blood' 115, **115**
Aubretia `Purple Cascade' 75, **75**
Aubriet, Claude 75
August 150–69
autumn crocus 191, **191**, 218, **218**
autumn daffodil 194, **194**
azalea (indoor) 236, **236**

backhouse Australian fuchsia 226, **226**
badgers 64
Balkan spurge 205, **205**
Baltic parsley 203, **203**
Banks, Sir Joseph 226
barberry 86, **86**
barrenwort 77, **77**
bearded iris (*Iris* `Sostenique') 92, **92**
bear's breeches 99, **99**, 141, **141**
Beaton, Donald 86
Beaton's currant *see* Gordon's currant
bees 13, 17, 26, 35, 39, 43, 56, 59, 65, 76, 79, 87–8, 98, 102, 104, 113, 115, 118, 126–7, 131, 135, 141, 145–6, 158, 166, 178, 206
 hairy-footed flower 73
 red mason 126
 solitary 135
 see also bumblebees
Begonia `Illumination Scarlet' 193, **193**
belladonna lily (*Amaryllis belladonna*) 198, 220, **220**
Bellis `Bellissima' 215, **215**
Berberis darwinnii `Compacta' 86, **86**
bergamot 178, **178**
Bergenia 17
 B. `Sunningdale' (elephant's ears) 59, **59**
Betula pendula 66, **66**
biennials 87, 102, 113, 126, 135, 159, 165
big blue lilyturf 196, **196**
bird cherry 104, **104**
birds 64, 82, 143, 164
 see also specific birds
bishop's flower 139, **139**
Bishop's hat *see* barrenwort
bistort, common 186, **186**
black-eyed Susan *see* coneflower
blackbirds 64, 89, 104, 158
blue wattle *see* mimosa (*Acacia dealbata*)
blue-grey hellebore 22, **22**
bluebeard 166, **166**
bluebells (*Hyacinthoides*) 55, 91, **91**, 92
Borago officinalis 199, **199**
Boronia crenulata 227, **227**
bowden lily (*Nerine bowdenii*) 177, **177**, 198
box 92
Brazilian fuchsia 208, **208**
Briza media 164, **164**
Buddleia globosa 146, **146**
Buelder, Jelena 18
Buelder, Robert 18
bulbs 151, 177, 189
 April-flowering 72–3, 75, 78, 81
 December-flowering 234–5, 237, 244, 246
 February-flowering 34, 35, 42, 43, 45, 46, 47, 48
 January-flowering 14, 17, 27–31
 July-flowering 132, 134
 March-flowering 51–5, 60, 62, 67–8
 May-flowering 91–3, 95
 November-flowering 219, 220, 222, 224
 October-flowering 191, 192, 194, 198
bullfinches 71
bumblebees 19, 35, 43–5, 106, 120, 219, 221, 238, 242
 buff-tailed 126
 red-tailed 126
Bupleurum rotundifolium (hare's ear) 173, **173**
buttercup witch hazel *see* winter hazel
butterflies 59, 76, 92, 98, 102, 115, 122, 127, 135, 145, 146, 158, 166, 206
butterfly amaryllis 237, **237**

cactus 60, 155, 233
Calendula 119, **119**
Calibrochoa 106, **106**
California poppy 138, **138**
Californian lilac *see* creeping blue blossom
Caltha palustris 65, **65**
Camellia 246
 C. sasanqua `Sparkling Beauty' 249, **249**
 C. transnokoensis 245, **245**
 C. x *veranlis* `Yuletide' 243, **243**
candelabra primula 75, **75**
candy cane sorrel 222, **222**
cannas 152
cape primrose 212, **212**
Caryopteris x *cladonensis* `Dark Knight' 166, **166**
caterpillars 59, 71, 81, 178
catkins 20, 39, 145
Ceanothus thyrsiflorus var. *repens* 80, **80**
Cenolophium denudatum 203, **203**
Centaurea montana (cornflower) 140, **140**, 147
Centranthus ruber 207, **207**
Ceratostigma plumbaginoides 179, **179**
Cercis chinensis `Avondale' 54, **54**
Cerinthe major `Purpurascens' 178, **178**
Chaenomeles japonica 79, **79**
cherry
 bird 104, **104**
 flagpole 82, **82**
 wild 64, **64**
 winter flowering 241, **241**
Chilean bellflower 229, **229**
Chilean iris 86, **86**
Chimonanthus praecox 20, **20**
Chinese redbud 54, **54**
Chinese wisteria 112, **112**
Chinese witch hazel 27
Chionodoxa luciliae 54, **54**
Chiron 140
chocolate vine 83, **83**
Christmas cactus 233, **233**

Christmas rose (*Helleborus niger*) 23, 24, **24**
Chrysanthemum 142, 198
 C. `Green Mist' 192, **192**
 C. segetum 148, **148**
Chrysosplenium macrophyllum 66, **66**
Cirsium heterophyllum 146, **146**
Clematis 37
 C. alpine 129
 C. `Christmas Surprise' 241, **241**
 C. cirrhosa 40
  C. c. var. *balearica* 23, **23**
  C. c. var. *purpurascens* `Freckles' 242, **242**
 C. `Madame Julia Correvon' 221, **221**
 C. napaulensis 238, **238**
 C. serratifolia 226, **226**
 C. tangutica `Bill Mackenzie' 212, **212**
 C. x *cartmanii* `Avalanche' 40
Cleome `Helen Campbell' 133, **133**
climbers 23, 40, 81, 83, 112, 131, 141, 212, 226, 238, 241
 see also deciduous climbers; evergreen climbers; perennial climbers
Cobaea scandens 187, **187**
cobnut *see* hazel
Colchicum `Waterlily' 191, **191**
Colletia paradoxa 225, **225**
colt's foot 33, **33**
Columbine *see* granny's bonnet
common bistort 186, **186**
common elder 123, **123**
common hogweed 135, **135**
cone bush 245, **245**
coneflower 186, **186**
Convallaria majalis (lily of the valley) 25, 74, **74**
Cootamundra wattle 248, **248**
coppicing 39
Coreopsis verticillate `Zagreb' 127, **127**
corms 14, 45, 157
corn cockle 147, **147**
corn marigold 148, **148**
corn poppy 148
corncockle 148
Cornelian cherry 21, **21**
cornflower (*Centaurea montana*) 140, **140**, 147
Cornus mas 21, **21**
Coronilla valentina supsp. *glauca* `Citrina' 34, **34**
Correa backhouseana 226, **226**
Corsican hellebore 58, **58**
Corylopsis pauciflora 84, **84**
Corylus avellana 39, **39**
Cosmos 133, 152, 172
 C. bipinnatus `Purity' 142, **142**
cow parsley 135, 139
Coward, Noel 155
cowslip (*Primula veris*) 76, **76**, 78
crab apple 66, **66**
Crambe
 C. cordifolia 129, **129**
 C. maritima 112, **112**
cranesbill 98, **98**
 dusky 107, **107**
Crataegus monogyna 81, **81**
creeping blue blossom 80, **80**
creeping forget me not 57, **57**
crimson flag 195, **195**
Crocosmia 118, 161
 C. `Paul's Best Yellow' 157, **157**
Crocus 42
 C. chrysanthus var. fuscotinctus 35, **35**
 C. sativus 219, **219**
 C. speciosus 218, **218**
 C. x *cultorum* `Jeanne D'Arc' 45, **45**
crown imperials 53, **53**
Culpeper, Nicholas 146
cuneate cinquefoil 175, **175**
cup and saucer vine 187, **187**
Cuphea ignea 181, **181**
cuttings 63
Cyclamen 37
 C. cilicium 213, **213**
 C. coum 14, **14**
  C. c. subsp. *coum f. pallidum* `Album' 28, **28**
 C. hederifolium 151, **151**
  C. h. var. *hederifolium* f. *albiflorum* 218, **218**
 C. persicum varieties 234, **234**

daffodil 37, 54
 early dwarf 34, **34**
 hoop petticoat 60, **60**
 jonquilla 27
 miniature 73
 `Paperwhite Ziva' 235, **235**
 `Tête-a-Tête' 52, **52**
 trumpet 27, **27**
Dahlia 118, 138, 142, 161, 168, 186, 195
 D. `Checkers' 172
 D. `Christopher Taylor' 152, **152**
 D. `David Howard' 172
 D. `Frigoulet' 200, **200**
 D. `Gallery Art Nouveau' 172, **172**
 D. `Moor Place' 152, **152**
 D. `Oakwood Goldcrest' 152, **152**
 D. `Primrose Dahlia' 172, **172**
 D. `Topmix Pink' 171, **171**
daisy Bellissima series 215, **215**
Danford, Mrs C.G. 48
Daphne
 D. mezereum 43, **43**
 D. odora 20, **20**
Darwin, Charles 86
David, Father Armand 81
December 230–49
deciduous climbers 124, 221
deciduous grasses 206
deciduous shrubs 106, 146, 205, 241
 April-flowering 71, 79, 86
 August-flowering 158, 166
 February-flowering 37–8, 40–1, 43
 January-flowering 17–21, 23, 27
 June-flowering 111, 121, 129
 March-flowering 65, 69
 November flowering 218, 225
deciduous trees 39–40, 61, 64, 66, 81, 88–9, 104, 108, 241
Delosperma cooperi `Jewel of Desert Peridot' 201, **201**
Delphinium nudicale `Redcap' 112, **112**
Deschampsia cespitosa 163, **163**
Deutzia x *rosea* 121, **121**
Dianthus (pinks) 207
 D. `Doris' 167, **167**
Diascia personata `Hopleys' 105, **105**
Dichromena colorata 215, **215**
Digitalis (foxglove) 107
 D. purpurea 113, **113**
 D. viridiflora 115, **115**
dog rose 158, **158**
dog tooth's violet 53, **53**
dogwood 48
dormice 39, 64
dusky cranesbill 107, **107**
dwarf red larkspur 112, **112**

early dwarf daffodil 34, **34**
early squill 42, **42**
early stachyurus 61, **61**
Echinacea 98
 E. purpurea 137, **137**
Echinops
 E. ritro 165, **165**
  E. r. `Veitch's Blue' 165
Echium
 E. pininana 159, **159**
 E. vulgare 126, **126**
Edgeworthia chrysantha 41, **41**
elder, common 123, **123**
elephant's ears 59, **59**
Elwes, Henry John 62
Epimedium 17, 58, 107
 E. lishichenii 77, **77**
Eranthis 42
 E. hyemalis 13, **13**
Erica (winter heather) 17
 E. carnea
  E. c. `Myretoun Ruby' 14, **14**
  E. c. `Vivelli' 37, **37**
 E. carnea alba `Springwood White' 26, **26**
 E. x *darleyensis* `Darley Dale' 48, **48**
 E. x *veitchii* `Exeter' 48, **48**
ericaceous plants 26, 37, 85, 227
 see also acidic soil
Erigeron karvinskianus 206, **206**
Eriostemon myoporoides 228, **228**
Eryngium martimum 105, **105**
Erysimum
 E. `Apricot Delight' 63, **63**
 E. `Bowles Mauve' 98, **98**, 232, **232**
Erythronium dens-canis 53, **53**
Eschscholzia californica 138, **138**
Eucomis
 E. autumnalis 134, **134**
 E. comosa `Sparkling Burgundy' 134
Euphorbia 105, 132
 E. characias subsp. *wulfenii* 78, **78**
 E. mellifera 202
 E. oblongata 205, **205**
 E. pulcherrima 235, **235**
evening primrose *see* ozark sundrops
evergreen climbers 81, 131, 184, 229, 242
evergreen ferns 25
evergreen perennials 22, 25, 27, 58, 66, 232
evergreen shrubs 20, 115, 156, 160
 April-flowering 80, 85–6
 December-flowering 231–2, 243, 245, 248–9
 February-flowering 34, 37, 44, 48
 January-flowering 14, 17, 26
 July-flowering 145–6
 November-flowering 212, 226, 228–9
 October-flowering 198, 201, 206, 208
evergreen trees 16, 49
Exochorda macrantha `The Bride' 129

feathertop 162, **162**
February 32–49
February daphne 43, **43**
fennel 138, **138**
fern-leaved clematis 23, **23**
ferns 25, 56, 57, 107, 132, 161, 174
Filipendula ulmaria 144, **144**
flagpole cherry 82, **82**
flamingo flower 238, **238**

flowering maple 198, **198**
Foencilum vulgare 138, **138**
forced hyacinth 235, **235**
forget me not 60, **60**, 91
  creeping 57, **57**
  water 128, **128**
Forrest, George 132
Forsythia x *intermedia* `Spectabilis' 37, **37**
Fortune, Robert 19, 84, 211
fortune saxifrage 211, **211**
foxglove (*Digitalis*) 107
  common 113, **113**
  green 115, **115**
Fritillaria
  F. imperialis 53
  F. meleagris **11**, 75, **75**
Fuchsia fulgens 178, **178**

Galanthus (snowdrop) 19, 37, 43, 48, 58
  G. elwesii 62, **62**
  G. nivalis 14, **14**
Gallium odoratum 101, **101**
Galtonia candicans 189, **189**
garden nasturtium 141, **141**
Garrya elliptica 20, **20**
Garzania `Tiger Stripes' 172, **172**
Gaura lindheimeri `The Bride' 176, **176**
gentian sage 138, **138**
Geranium 154
  G. phaeum 107, **107**
  G. pratense `Mrs Kendall Clark' 98, **98**
  hardy 92, 116, 146, 192, 204
giant golden saxifrage 66, **66**
giant hogweed 135
giant viper's bugloss 159, **159**
giant wake robin 65, **65**
Gladiolus murielae 192
Glandora prostrata see Lithodora diffusa `Heavenly Blue'
Glebionsis segetum see Chrystanthemum segetum
globeflower 126, **126**
glory bush 247, **247**
glory of the snow 54, **54**
glossy abelia 186, **186**
golden clematis 212, **212**
golden mimosa *see* Cootamundra wattle
golden rain tree *see Laburnum anagyroides*
goldenrod 175, **175**
Gordon, William 86
Gordon's currant 86, **86**
granny's bonnet 94, **94**
grape hyacinth (*Muscari*) 25, 51, **51**, 52, 73, **73**
grasses (ornamental) 43, 105, 127, 132, 152, 161–5, 172, 176, 185, 195, 204, 206, 223
grassland 76
greater periwinkle 181, **181**
greater sea kale 129, **129**
greater snowdrop 62, **62**
green manure 201
greenfinches 66, 164
Grevillea rosmarinifolia 229, **229**
guelder rose 71, **71**
gypsophila 207

Hamamelis 37
  H. mollis `Brevipetala' 40, **40**
  H. mollis `Goldcrest' 69, **69**
  H. x *intermedia*
    H. x *i.* `Diane' 27, **27**
    H. x *i.* `Jelena' 18, **18**
    H. x *i.* `Strawberries and Cream' 38, **38**
hardy geranium 92, 116, 146, 192, 204
hardy plumbago 179, **179**
hare's ear 173, **173**
Hattie's pincushion *see* masterwort
hawthorn 81, **81**
hazel 39, **39**
heart's ease 54, **54**
heath aster 195, **195**
hebe 92
hedgehogs 158
Helenium 168, 192, 202
  H. `Moerheim Beauty' 118, **118**
Helianthus
  H. annuus 143, **143**, 161
  H. `Lemon Queen' 161, **161**
Heliotropium arborescens 122, **122**
Helleborus 19, 37, 48, 52, 56
  H. argutifolius 58, **58**
  H. `Ashwood Garden Hybrids' 34, **34**
  H. foetidus 25, **25**
  H. lividius 22, **22**
  H. niger (Christmas rose) 23, 24, **24**
  H. `Snow Dance' 246, **246**
  H. `Walburton's Rosemary' 23, **23**
  H. x *hybridus* 23, 34
Heptacodium miconioides 183, **183**, 216, **216**
Heracleum sphondylium 135, **135**
herbaceous peony 96, **96**
Hesperantha coccinea `Jennifer' 195, **195**
Hesperis matronalis 102, **102**
heuchera 56
Hibiscus syriacus `Woodbridge' 205, **205**
Himalayan fairy grass 223, **223**
Hippeastrum
  H. `Green Valley' 29, **29**
  H. papilio 237, **237**
  H. `Spartacus' 244, **244**
  H. (Spider Group) `Evergreen' 28, **28**
holly leaved hellebore *see* Corsican hellebore
hollyhock 136, **136**
honesty 87, **87**
honeybees 126
honeywort 178, **178**
hoop petticoat daffodil 60, **60**
hops 145, **145**
hostas 56, 132, 161, 174
house plants 233, 238, 239
house sparrows 164
hoverflies 71, 105, 145
Hoya carnosa 184, **184**
Humulus lupulus 145, **145**
Hyacinthoides non-scripta (bluebells) 55, 91, **91**, 92
Hyacinthus 235, **235**
Hydrangea
  H. arborescens `Annabelle' 121, **121**
  H. `Great Star' 155, **155**
Hylotelephium spectabile 154, **154**
Hypericum
  H. perforatum 206
  H. x *hidcoteense* `Hidcote' 206, **206**
Hyssopus officinalis 188, **188**

ice plant (*Delosperma cooperi* `Jewel of Desert Peridot') 201, **201**
iceplant (*Hylotelephium spectabile*) 154, **154**
`in the green' 13, 62, 91
indoor cyclamen 234, **234**
Ipomoea lobata 197, **197**
Iris
  I. danfordiae 48, **48**
  I. `Mary Barnard' 246, **246**
  I. reticulata `George' 47, **47**
  I. reticulata `Katherine Hodgkin' 46, **46**, 47
  I. `Sostenique' 92, **92**
  I. unguicularis `Mary Barnard' 43
  I. unguicularis `Walter Butt' 43
  miniature 54
Isopogon formosus 245, **245**
ivy 52
ivy-leaved cyclamen (*Cyclamen hederifolium*) 151, **151**, 218, **218**

January 12–31
Japanese anemone 161, **161**, 180, **180**
Japanese mahonia 232, **232**
Japanese quince 79, **79**
Japanese witch hazel 27
Jasminum
  J. nudiflorum 23, **23**
  J. officinale 124, **124**
Jekyll, Gertrude 129
Jersey lily *see* belladonna lily
Johnston, Laurence 145
jonquilla daffodils 27
July 130–49
June 110–29
Justicia floribunda rizzinii 208, **208**

Kirengoshoma palmata 174, **174**
Kniphofia (red-hot poker) 78
  K. `Ice Queen' 202, **202**
  K. uvaria 161, **161**
Korean clematis 226, **226**

Laburnum anagyroides 108, **108**
ladybirds 163
Lantana camara 145, **145**
Lapageria rosea 229, **229**
Lathyrus odoratus 101, **101**
Lavandula angustifolia `Hidcote' 145, **145**
Lent lily *see* wild daffodil
Leonotis leonurus 198, **198**
lesser celandine 43, **43**
lesser periwinkle 181
Leucanthemum vulgare 149, **149**
Libertia formosa 86, **86**
lilac 106, **106**
Lilium candidum 158, **158**
lily of the valley (*Convallaria majalis*) 25, 74, **74**
linnets 164
lion's tail 198, **198**
Liriope muscari 196, **196**
Lithodora diffusa `Heavenly Blue' 156, **156**
Lloyd, Christopher 221
Lobb, William 229
Lobularia maritima `Snowdrift' 125, **125**
long leaf wax flower 228, **228**
Lonicera
  L. fragrantissima 19, **19**
  L. x *purpusii* `Winter Beauty' 238, **238**
Lorapetalum chinense var. *rubrum* `Fire Dance' 37, **37**
Louis XIV 75
love-in-a-mist 101, **101**
Lunaria annua 87, **87**
lungwort 49, **49**
Lupinus arboreus 115, **115**
Lychnis coronaria 204, **204**

Madeiran squill 30, **30**
Madonna lily 158, **158**
Magnolia stellata 65, **65**
Mahonia

M. japonica 232, **232**
M. nitens 'Cabaret' 212, **212**
M. x *media* 44, **44**
M. x *m.* 'Winter Sun' 17, **17**
Malus sylvestris 66, **66**
Malva moschata 166, **166**
Mammillaria spinosissima 60, **60**
March 50–69
marginal pond plants 65, 128, 215
marguerite 226, **226**
marsh marigold 65, **65**
Masson's wood sorel 224, **224**
masterwort 115, **115**
Mathiasella bupleuroides 'Green Dream' 95, **95**
Matthiola longipetala 97, **97**, 158, **158**
meadowsweet 144, **144**
medicinal plants 144, 146, 165, 206, 215, 245
Mediterranean spurge 78, **78**
melancholy thistle 146, **146**
Mexican cigar plant 181, **181**
Mexican fleabane 206, **206**
Mexican giant hyssop 192, **192**
Mexican sunflower 168, **168**
mice 39, 64
Michaelmas daisy 175, **175**
Middendorf weigela 111, **111**
million bells 106, **106**
mimosa (*Acacia dealbata*) 16, **16**
Mimosa pudica (sensitive plant) 205, **205**
Miscanthus
M. nepalensis 223, **223**
M. sacchariflorus 206, **206**
M. sisensis 'Zebrincus' 185, **185**
mock orange 121, **121**
Monarda didyma 'Cambridge Scarlet' 178, **178**
montbretia 157, **157**
moth orchid 239, **239**
moths 59, 81, 92, 97, 102, 115, 144, 158
Mount Noko camellia 245, **245**
mountain ash *see* rowan
Muscari 25, 52
M. armeniacum 51, 73, **73**
M. latifolium 51, **51**
musk mallow 166, **166**
Myosotis
M. scorpioides 128, **128**
M. sylvatica 60, **60**
Myrtus communis 146, **146**

Narcissus 43
dwarf 25
N. bulbocodium 60, **60**
N. 'February Gold' 34, **34**
N. papyraceus 'Paperwhite Ziva' 235, **235**
N. poeticus 78, **78**
N. pseudonarcissus 68, **68**
N. 'Rijnveld's Early Sensation' 27, **27**
N. 'Tête-a-Tête' 52, **52**
nasturtium, garden 141, **141**
native primrose 59, **59**
Nectaroscordum siculum 95, **95**
Nepeta 105, 118, 127
N. subsessilis 132, **132**
Nerine bowdenii (bowden lily) 177, **177**, 198
newts 128
nicotiana 172
Nigella damascena 101, **101**
night-scented stock 97, **97**, 158, **158**
nitrogen fixation 201
November 210–29

Nymphaea odorata subsp. *tuberosa* 153, **153**

October 190–209
Oenothera macrocarpa 165, **165**
Omphalodes verna 57, **57**
Ononis spinosa 169, **169**
opium poppy 141, **141**
orange ball tree 146, **146**
orchids, moth 239, **239**
Oregon grape 17, **17**, 44, **44**, 212, **212**
oriental poppy 95, **95**
Orlaya grandiflora 105, **105**
ornamental onion 92, **92**
ornamental pear 88, **88**
Osmanthus fragrans f. *aurantiacus* 209, **209**
ox-eye daisy 149, **149**
Oxalis
O. massoniana 224, **224**
O. versicolor 222, **222**
oxlip 78, **78**
ozark sundrops 165, **165**

pansy (*Viola* x *wittrockiana*) 15, **15**, 73
'Frizzle Sizzle Burgundy' 17, **17**
'Mystique Blue Whiskers' 28, **28**
Papaver
P. orientale 95, **95**
P. rhoeas 120, **120**
P. somniferum 141, **141**
paperbush 41, **41**
Parrotia persica 40, **40**
Passiflora caerulea (passionflower) 131, **131**
pearl bush 129
Pelargonium
P. 'Lord Bute' 214, **214**
P. sidoides 215, **215**
Pennisetum
P. 'Red Head' 172
P. villosum 162, **162**
Penstemon
P. 'Geoff Hamilton' 135, **135**
P. 'Raven' 221, **221**
Peonia
P. lactiflora 'Bowl of Beauty' 96, **96**
P. x *suffruticosa* 114, **114**
peony
herbaceous 96, **96**
tree 114, **114**
perennial climbers 145, 187
perennial phlox 181, **181**
perennial wallflower (*Erysimum*)
'Apricot Delight' 63, **63**
'Bowles' Mauve' 98, **98**, 232, **232**
perennials
April-flowering 72, 74–8, 86
August-flowering 152, 154–5, 157–8, 161–7, 169
December-flowering 232–3, 246–7
February-flowering 33–4, 36, 43, 49
July-flowering 132, 135–8, 140–1, 144, 146, 149
June-flowering 112, 115–18, 126–7, 129
March-flowering 56–60, 63, 65
May-flowering 94–6, 98–101, 103, 105–7
November-flowering 211–15, 217, 221, 223, 226
October-flowering 192, 195–6, 200–7
September-flowering 171–2, 174–6, 178–82, 185–6, 188–9
short-lived 106
see also evergreen perennials; semi evergreen perennials
Persian ironwood 40, **40**

Persicaria bistorta 'Superba' 186, **186**
Peruvian lily 166, **166**, 221, **221**
Phalaenopsis varieties 239, **239**
pheasant's eye daffodil 78, **78**
Philadelphus 'White Rock' 121, **121**
Phlomis russeliana 165, **165**
Phlox paniculata 181, **181**
Pieris japonica 'Purity' 85, **85**
pineapple lily 134, **134**
pinks (*Dianthus*) 207
D. 'Doris' 167, **167**
Plantlife (charity) 76
poinsettia (*Euphoriba pulcherrima*) 235, **235**
Polygonatum x *hybridum* 100, **100**
Polyspora speciosa 246, **246**
poppy 147
California 138, **138**
field 120, **120**
opium 141, **141**
oriental 95, **95**
Portuguese tree heather 48, **48**
Portulaca umbraticola 109, **109**
pot marigold 119, **119**
Potentilla
P. cuneata 175, **175**
P. fruticosa 'Elizabeth' 195, **195**
Primula
P. allionii 75, **75**
P. elatior 78, **78**
P. veris 76, **76**, 78
P. vialii 132
P. vulgaris 59, **59**
Prunus
P. avium 64, **64**
P. domestica 'Victoria' 89, **89**
P. padus 104, **104**
P. x *subhirtella* 'Autumnalis Rosea' 241, **241**
P. x *yoedensis* 82, **82**
Pulmonaria **7**, 57
P. rubra 56, **56**
purple coneflower 137, **137**
purple gromwell 156, **156**
Puschkinia scilloides var. *libanotica* 72, **72**
Pyrus calleryana 'Chanticleer' 88, **88**

quaking grass 164, **164**

Radde's fritillary 53
red clover 201, **201**
red lungwort 56, **56**
red mason bees 126
red valerian 207, **207**
red-hot poker (*Kniphofia*) 78, 161, **161**
'Ice Queen' 202, **202**
redwings 158
regal pelargonium 214, **214**
restharrow 169, **169**
reticulata iris 48, **48**
'George' 47, **47**
'Katherine Hodgkin' 46, **46**, 47
rhizomes 33, 55, 72, 92
Rhododendron simsii varieties 236, **236**
Ribes x *beatonii* 86, **86**
rodgersia 174
Rosa (rose) 36
R. canina (dog rose) 158, **158**
rose campion 204, **204**
rose deutzia 121, **121**
Rose of Sharon *see* tree hollyhock
rosemary grevillea 229, **229**
rowan 89, **89**
Royal Horticultural Society (RHS)

Award of Garden Merit (AGM) 13
Plants for Pollinators 26
Rudbeckia 118, 154, 163, 195, 202
R. fulgida var. *sullivantii* `Goldsturm' 186, **186**
rue anemone 59, **59**
Russian snowdrop *see* striped squills

saffron crocus 219, **219**
sage *see Salvia*
St John's Wort 206, **206**
Salvia 105, 127
S. `Amistad' 118, **118**
S. gauranitica `Black and Blue' 182, **182**
S. `Jezebel' 103, **103**
S. nemorosa `Caradonna' 116, **116**
S. patens 138, **138**
Sambucus nigra 123, **123**
sanguisorba 195
Sarcococca confusa 232, **232**
Saxifraga
S. fortunei `Silver Velvet' 211
S. spathularis 155
S. umbrosa 155, **155**
S. x *urbium* 155
Scabiosa 207
S. columbaria 118, **118**
Schlumbergera 233, **233**
Scilla
S. madeirensis 30, **30**
S. mischtschenkoana 42, **42**
S. siberica 53
scorpion vetch 34, **34**
sea holly 105, **105**
sea kale 112, **112**
sea thrift 98, **98**
Sedum 154
S. brevifolium 155, **155**
semi evergreen perennials 23, 24
semi evergreen shrubs 198
sensitive plant (*Mimosa pudica*) 205, **205**
September 170–89
seven son flower tree 183, **183**, 216, **216**
short leaved stonecrop 155, **155**
short-stalked catmint 132, **132**
shrubby acalthea `Parkalee' 217, **217**
shrubby cinquefoil 195, **195**
shrubs 78, 84, 114, 122, 123, 155, 178, 181, 183, 186, 195, 216, 227, 236, 238, 246
see also deciduous shrubs; evergreen shrubs; semi evergreen shrubs
Siberian squill 53
Sicilian honey garlic 95, **95**
silver banner grass 206, **206**
silver birch 66, **66**
silver tassel bush 20, **20**
siskins 66
skimmia 17
small globe thistle 165, **165**
small leaved kowhai 49, **49**
small tree 246
snake's head fritillary **11**, 75, **75**
snapdragon 106, **106**
sneezeweed 118, **118**
snowdrop (*Galanthus*) 14, **14**, 19, 37, 43, 48, 58
greater 62, **62**
snowflake 67, **67**
Soldago rugosa `Fireworks' 175, **175**
Solomon's seal 100, **100**
Sophora microphylla 49, **49**
Sorbus aucuparia 89, **89**
sowbread 14, **14**, 28, **28**, 213, **213**, 218, **218**
Spanish bluebell 91
Spanish flag 197, **197**
speedwell 189, **189**
spider flower 133, **133**
spiny pin cushion cactus 60, **60**
squirrels 39
stachys 105, 116, 154
Stachyurus praecox 61, **61**
star grass 215, **215**
star magnolia 65, **65**
star of Persia 132, **132**
Sternberg, Count Kaspar Moritz von 194
Sternbergia lutea 194, **194**
stinking hellebore 25, **25**
Stipa
S. gigantea 172
S. tenuissima 127
strawberry tree 201, **201**
Streptocarpus `Falling Stars' 212, **212**
striped squills 72, **72**
succulents 109, 155
summer hyacinth 189, **189**
sunflower 143, **143**, 161, **161**
sweet alyssum 125, **125**
sweet box 232, **232**
sweet pea 101, **101**
sweet rocket 102, **102**
sweet tea 209, **209**
sweet violet 36, **36**
sweet woodruff 101, **101**
Symphyotrichum
S. ericoides `Pink Cloud' 195, **195**
S. `Little Carlow' 175, **175**
Syringa vulgaris 106, **106**

tadpoles 128
Thalictrum 59
T. thalictroides 59, **59**
thrush 104
mistle 71, 89
song 64, 89
Tibouchina paratropica 247, **247**
tickseed 127, **127**
Tithonia rotundifolia 168, **168**
tree heath 48
tree hollyhock 205, **205**
tree lupin 115, **115**
tree peony 114, **114**
trees 54, 82, 123
deciduous 39–40, 61, 64, 66, 81, 88–9, 104, 108, 241
evergreen 16, 49
Trifolium pratense 201, **201**
Trillium chloropetalum 65, **65**
Tristram, David 23
Tristram, Rosemary 23
Trollius europaeus 126, **126**
Tropaeolum majus 141, **141**
true London pride 155, **155**
trumpet daffodil 27, **27**
tubers 13, 28, 43, 151, 152, 171
tufted hair grass 163, **163**
Tulipa 42, 73, 92, 98, 192
T. `Oviedo' 92, **92**
T. sylvestris 72, **72**
T. `Typhoon' 81, **81**
Turkish sage 165, **165**
Tussilago farfara 33, **33**
twinspur 105, **105**

verbascum 105, 204
Verbena 78, 165
V. bonariensis (Argentinian vervain) 135, **135**, 152, 172, 175, **175**
Veronica
V. longifolia `Marietta' 189, **189**
V. x *andersonii* 160, **160**
vial's primrose 132
Viburnum
V. opulus 71, **71**
V. tinus 231, **231**
V. x *bodnantense*
V. x *b.* `Charles Lamont' 218, **218**
V. x *b.* `Dawn' 241, **241**
Victoria plum 89, **89**
Vinca major 181, **181**
Viola
V. odorata 36, **36**
V. `Tiger Eyes' 27, **27**
V. tricolor 54, **54**
V. x *wittrockiana* 15, **15**, 73
V. x *w.* `Frizzle Sizzle Burgundy' 17, **17**
V. x *w.* `Mystique Blue Whiskers' 28, **28**
viper's bugloss 126, **126**
see also giant viper's bugloss
von Alströmer, Baron 166
von Siebold, Philipp Franz 61

Walburton's Nursery 23
water forget me not 128, **128**
water hawthorn 240, **240**
waterlily 153, **153**
wax plant 184, **184**
Weigela middendorffiana 111, **111**
white forsythia 17, **17**
white laceflower 105, **105**
wild cherry 64, **64**
wild daffodil 68, **68**
wild hyacinth **93**
wild tulip 72, **72**
Wildlife Trust, The 104
Wilson, EH 226
wingpod purslane 109, **109**
winter aconite 13, **13**
winter daphne 20, **20**
winter flowering cherry 241, **241**
winter flowering clematis 238, **238**, 241, **241**, 242, **242**
winter hazel 84, **84**
winter heather (*Erica*) 14, **14**, 17, 26, **26**, 37, **37**, 48, **48**
winter honeysuckle 19, **19**, 238, **238**
winter jasmine 23, **23**
wintersweet 20, **20**
Wisteria sinensis 112, **112**
witch hazel (*Hamamelis*) 18, **18**, 27, **27**, 37–8, **38**, 40, **40**, 69, **69**
Wittrock, Veit Brecher 15
wood anemone 55, **55**, 72, **72**
wood pigeon 39
woodland plants 13
Woodland Trust, The 55, 78
woodpeckers 39

yarrow 117, **117**
yellow sage 145, **145**
yellow wax bells 174, **174**

zebra grass 185, **185**
Zuccarni, Joseph 61Text ...Text ...

PICTURE CREDITS

Every effort has been made to locate and credit copyright holders of the material reproduced in this book. The publisher apologise for any omissions or errors, which can be corrected in future editions. a=above, c=centre, b=below, BHL= Biodiversity Heritage Library

2 The Natural History Museum/Alamy Stock Photo; 7 Image from the BHL. Contributed by The LuEsther T Mertz Library, the New York Botanical Garden; 8 Image from the BHL. Contributed by New York Botanical Garden; 11 Image from the BHL. Contributed by Lloyd Library and Museum; 13 Jason Ingram © Gardeners' World Magazine; 14a Image from the BHL. Contributed by Chicago Botanic Garden, Lenhardt Library; 14c Image from the BHL. Contributed by New York Botanical Garden, LuEsther T. Mertz Library; 14b fotorince; 15 Image from the BHL. Contributed by Missouri Botanical Garden; 16 Image from the BHL. Contributed by Smithsonian Libraries and Archives; 17a N.Stertz; 17c KUMA-MON; 17b, 18 Jason Ingram © Gardeners' World Magazine; 19 Image from the BHL. Contributed by Missouri Botanical Garden; 20a user685475/Alamy Stock Vector; 20c Image from the BHL. Contributed by Missouri Botanical Garden; 20b Library of Congress, Washington D.C.; 21 Quagga Media/Alamy Stock Photo; 22 Image from the BHL. Contributed by California Academy of Sciences; 23a The Natural History Museum/Alamy Stock Photo; 23c Paul Debois © Gardeners' World Magazine; 23b Sarah Cuttle © Gardeners' World Magazine; 24 Image from the BHL. Contributed by Lloyd Library and Museum; 25 Image from the BHL. Contributed by New York Botanical Garden, LuEsther T. Mertz Library; 26, 27a Jason Ingram © Gardeners' World Magazine; 27c Tim Sandall © Gardeners' World Magazine; 27b, 28a Jason Ingram © Gardeners' World Magazine; 28c imageBROKER/Shutterstock; 28b visitfaerie/Alamy Stock Photo; 29 Sarah Cuttle © Gardeners' World Magazine; 30 juerginho; 31 Sarycheva Olesia; 33 Image from the BHL. Contributed by Field Museum of Natural History Library; 34a, 34c Jason Ingram © Gardeners' World Magazine; 34b, 35 Image from the BHL. Contributed by Missouri Botanical Garden; 36 Image from the BHL. Contributed by Smithsonian Libraries and Archives; 37a Torie Chugg © Gardeners' World Magazine; 37c Anna Gratys; 37b Nahhana; 38 Jason Ingram © Gardeners' World Magazine; 39 Image from the BHL. Contributed by Chicago Botanic Garden, Lenhardt Library; 40a Peter Turner Photography; 40c Jason Ingram © Gardeners' World Magazine; 40b Jose Luis Vega; 41 Image from the BHL. Contributed by Missouri Botanical Garden, Peter H. Raven Library; 42 Nick Pecker; 43a Quagga Media/Alamy Stock Photo; 43c Image from the BHL. Contributed by New York Botanical Garden, LuEsther T. Mertz Library; 43b Zuri Swimmer/Alamy Stock Photo; 44 Jason Ingram © Gardeners' World Magazine; 45 From sky to the Earth; 46 Sarah Cuttle © Gardeners' World Magazine; 47 Jason Ingram © Gardeners' World Magazine; 48a, 48c Sarah Cuttle © Gardeners' World Magazine; 48b Nahhana; 49a The Natural History Museum/Alamy Stock Photo; 49b Image from the BHL. Contributed by New York Botanical Garden, LuEsther T. Mertz Library; 51, 52, 53a, 53c Image from the BHL. Contributed by Missouri Botanical Garden, Peter H. Raven Library; 53b Image from the BHL. Contributed by New York Botanical Garden, LuEsther T. Mertz Library; 54a Traveller70; 54c Image from the BHL. Contributed by Missouri Botanical Garden, Peter H. Raven Library; 54b Image from the BHL. Contributed by New York Botanical Garden, LuEsther T. Mertz Library; 55 Image from the BHL. Contributed by University of California Libraries; 56 Image from the BHL. Contributed by New York Botanical Garden, LuEsther T. Mertz Library; 57 Image from the BHL. Contributed by Missouri Botanical Garden, Peter H. Raven Library; 58 Chris Lawrence Images; 59a Image from the BHL. Contributed by New York Botanical Garden, LuEsther T. Mertz Library; 59c Jason Ingram © Gardeners' World Magazine; 59b Image from the BHL. Contributed by Missouri Botanical Garden, Peter H. Raven Library; 60a The History Collection/Alamy Stock Photo; 60c Image from the BHL. Contributed by New York Botanical Garden, LuEsther T. Mertz Library; 60b Image from the BHL. Contributed by Missouri Botanical Garden, Peter H. Raven Library; 61 tamu1500; 62 Image from the BHL. Contributed by Missouri Botanical Garden, Peter H. Raven Library; 63 Jason Ingram © Gardeners' World Magazine; 64 Florilegius/Alamy Stock Photo; 65a The Garden: an illustrated weekly journal of gardening in all its branches, vol.13, William Robinson, 1878; 65c Smithsonian American Art Museum, Gift of the artist, 1970.355.636; 65b, 66a Image from the BHL. Contributed by Missouri Botanical Garden, Peter H. Raven Library; 66c Courtesy of the Smithsonian Libraries and Archives; 66b Tintila Corina; 67, 68 Image from the BHL. Contributed by Missouri Botanical Garden, Peter H. Raven Library; 69 Jason Ingram © Gardeners' World Magazine; 71 Image from the BHL. Contributed by New York Botanical Garden, LuEsther T. Mertz Library; 72a, 72c Image from the BHL. Contributed by Missouri Botanical Garden, Peter H. Raven Library; 72b Walter Erhardt; 73 Old Images/Alamy Stock Photo; 74 Image from the BHL. Contributed by Missouri Botanical Garden, Peter H. Raven Library; 75a Image from the BHL. Contributed by Lloyd Library and Museum; 75c Tim Gainey/Alamy Stock Photo; 75b Antoniya Kadiyska; 76 Image from the BHL. Contributed by New York Botanical Garden, LuEsther T. Mertz Library; 77 Torie Chugg © Gardeners' World Magazine; 78a Image from the BHL. Contributed by Missouri Botanical Garden, Peter H. Raven Library; 78c Wellcome Collection, London; 78b Jason Ingram © Gardeners' World Magazine; 79 Image from the BHL. Contributed by Lloyd Library and Museum; 80 Image from the BHL. Contributed by Missouri Botanical Garden, Peter H. Raven Library; 81a Sergey V Kalyakin; 82c Image from the BHL. Contributed by Missouri Botanical Garden, Peter H. Raven Library; 81b Basicmoments/Alamy Stock Photo; 82 N.Stertz; 83 Image from the BHL. Contributed by Missouri Botanical Garden, Peter H. Raven Library; 84 Paul Debois © Gardeners' World Magazine; 85, 86a Jason Ingram © Gardeners' World Magazine; 86c Florilegius/Alamy Stock Photo; 86b Sarah Cuttle © Gardeners' World Magazine; 87 Image from the BHL. Contributed by Lloyd Library and Museum; 88 LUXURIOUS.thelabel; 89a markku murto/art/Alamy Stock Photo; 89b Image from the BHL. Contributed by University of Toronto - Robarts Library; 91 Image from the BHL. Contributed by MBLWHOI Library; 92a Angiolino Baruffa; 92c Holmes Garden Photos/Alamy Stock Photo; 92b Sarah Cuttle © Gardeners' World Magazine; 93 Paul Debois © Gardeners' World Magazine; 94 Image from the BHL. Contributed by New York Botanical Garden, LuEsther T. Mertz Library; 95a Torie Chugg © Gardeners' World Magazine; 95c Image from the BHL. Contributed by Missouri Botanical Garden, Peter H. Raven Library; 95b Sarah Cuttle © Gardeners' World Magazine; 96 Jason Ingram © Gardeners' World Magazine; 97 Image from the BHL. Contributed by New York Botanical Garden, LuEsther T. Mertz Library; 98a Sarah Cuttle © Gardeners' World Magazine; 98c Image from the BHL. Contributed by Missouri Botanical Garden, Peter H. Raven Library; 98b Katherine Hugh uk; 99 Image from the BHL. Contributed by Lloyd Library and Museum; 100 Jason Ingram © Gardeners' World Magazine; 101a Image from the BHL. Contributed by University of California Libraries; 101c, 102 Image from the BHL. Contributed by Missouri Botanical Garden, Peter H. Raven Library; 101b Image from the BHL. Contributed by New York Botanical Garden, LuEsther T. Mertz Library; 103 Paul Debois © Gardeners' World Magazine; 104 Image from the BHL. Contributed by New York Botanical Garden, LuEsther T. Mertz Library; 105a Wellcome Collection, London, 105c,105b Sarah Cuttle © Gardeners' World Magazine; 106a Image from the BHL. Contributed by Missouri Botanical Garden, Peter H. Raven Library; 106c Florilegius/Alamy Stock Photo; 106b Alina Kuptsova; 107 Image from the BHL. Contributed by New York Botanical Garden, LuEsther T. Mertz Library; 108 Image from the BHL. Contributed by Field Museum of Natural History Library; 109 Anggoro21; 111 Image from the BHL. Contributed by Smithsonian Libraries and Archives; 112a Image from the BHL. Contributed by Missouri Botanical Garden, Peter H. Raven Library; 112c Jason Ingram © Gardeners' World Magazine; 112b, 113 Image from the BHL. Contributed by New York Botanical Garden, LuEsther T. Mertz Library; 114 Image from the BHL. Contributed by Lloyd Library and Museum; 115a Justin Lambert © Gardeners' World Magazine; 115c The Natural History Museum/Alamy Stock Photo; 115b Image from the BHL. Contributed by Missouri Botanical Garden, Peter H. Raven Library; 116 Jason Ingram © Gardeners' World Magazine; 117 American medicinal plants; an illustrated and descriptive guide to the American plants used as homopathic remedies, Vol.1, Charles F. Millspaugh, 1887; 118a Jason Ingram © Gardeners' World Magazine; 118c Torie Chugg © Gardeners' World Magazine; 118b Image from the BHL. Contributed by New York Botanical Garden, LuEsther T. Mertz Library; 119 Image from the BHL. Contributed by Missouri Botanical Garden, Peter H. Raven Library; 120 Image from the BHL. Contributed by Field Museum of Natural History Library; 121a, 121c, 121b Jason Ingram © Gardeners' World Magazine; 122 Image from the BHL. Contributed by Missouri Botanical Garden, Peter H. Raven Library; 123 Wellcome Collection, London; 124 Image from the BHL. Contributed by Field Museum of Natural History Library; 125 Rose Marinelli; 126a Image from the BHL. Contributed by New York Botanical Garden, LuEsther T. Mertz Library; 126c Wellcome Collection, London; 126b Image from the BHL. Contributed by New York Botanical Garden, LuEsther T. Mertz Library; 127 Sarah Cuttle © Gardeners' World Magazine; 128 Image from the BHL. Contributed by Missouri Botanical Garden, Peter H. Raven Library; 129a Clare Gainey/Alamy Stock Photo; 129b Paul Debois © Gardeners' World Magazine; 131 Image from the BHL. Contributed by New York Botanical Garden, LuEsther T. Mertz Library; 132a Image from the BHL. Contributed by Missouri Botanical Garden, Peter H. Raven Library; 132c Sarah Cuttle © Gardeners' World Magazine; 132b, 134 Image from the BHL. Contributed by Missouri Botanical Garden, Peter H. Raven Library; 133 Old Images/Alamy Stock Photo; 135a Paul Debois © Gardeners' World Magazine; 135c Image from the BHL. Contributed by New York Botanical Garden, LuEsther T. Mertz Library; 135b Old Images/Alamy Stock Photo; 136 Paul Debois © Gardeners' World Magazine; 137, 138a Image from the BHL. Contributed by Missouri Botanical Garden, Peter H. Raven Library; 138c, 138b Image from the BHL. Contributed by New York Botanical Garden, LuEsther T. Mertz Library; 139 New York Public Library, New York; 140, 141c Image from the BHL. Contributed by Missouri Botanical Garden, Peter H. Raven Library; 141a New York Public Library, New York; 141b Wellcome Collection, London; 142 Paul Debois © Gardeners' World Magazine; 143 The Natural History Museum/Alamy Stock Photo; 144, 145c Image from the BHL. Contributed by New York Botanical Garden, LuEsther T. Mertz Library; 145a Image from the BHL. Contributed by Missouri Botanical Garden, Peter H. Raven Library; 145b Jason Ingram © Gardeners' World Magazine; 146a Wellcome Collection, London; 146c Album/Alamy Stock Photo; 146b Image from the BHL. Contributed by Missouri Botanical Garden, Peter H. Raven Library; 147 Image from the BHL. Contributed by New York Botanical Garden, LuEsther T. Mertz Library; 148, 149 Wellcome Collection, London; 151 Image from the BHL. Contributed by Lloyd Library and Museum; 152a Sarah Cuttle © Gardeners' World Magazine; 152c Stephen William Robinson; 152b Torie Chugg © Gardeners' World Magazine; 153 Martin Fowler; 154 Image from the BHL. Contributed by Missouri Botanical Garden, Peter H. Raven Library; 155a Wellcome Collection, London; 155c, 156, 157 Paul Debois © Gardeners' World Magazine; 155b Sarah Cuttle © Gardeners' World Magazine; 158a Image from the BHL. Contributed by Missouri Botanical Garden, Peter H. Raven Library; 158c Image from the BHL. Contributed by University of California Libraries; 158b, 159 Jason Ingram © Gardeners' World Magazine; 160 Image from the BHL. Contributed by Missouri Botanical Garden, Peter H. Raven Library; 161a, 161b, 162 Torie Chugg © Gardeners' World Magazine; 161c, 164, 165a, 165b Image from the BHL. Contributed by Missouri Botanical Garden, Peter H. Raven Library; 163 Wellcome Collection, London; 165c Image from the BHL. Contributed by New York Botanical Garden, LuEsther T. Mertz Library; 166a Image from the BHL. Contributed by University of California Libraries; 166c Hans-Roland Mueller/imageBROKER/Shutterstock; 166b Paul Debois © Gardeners' World Magazine; 167 Jason Ingram © Gardeners' World Magazine; 168 Image from the BHL. Contributed by Missouri Botanical Garden, Peter H. Raven Library; 169 Image from the BHL. Contributed by New York Botanical Garden, LuEsther T. Mertz Library; 171 thrillerfillerspiller/Alamy Stock Photo; 172a keith morris/Alamy Stock Photo; 172c, 172b, 174 Sarah Cuttle © Gardeners' World Magazine; 173 Image from the BHL. Contributed by Harvard University Botany Libraries; 175a Bremar; 175c tamu1500; 175b Traveller70; 176 Landrausch; 177, 178c, 179 Image from the BHL. Contributed by Missouri Botanical Garden, Peter H. Raven Library; 178a Jason Ingram © Gardeners' World Magazine; 178b Image from the BHL. Contributed by New York Botanical Garden, LuEsther T. Mertz Library; 180 Peter Turner Photography; 181a Image from the BHL. Contributed by University of California Libraries; 181c, 181b Image from the BHL. Contributed by Missouri Botanical Garden, Peter H. Raven Library; 182 Nikolay Kurzenko; 183, 185, 186a Jason Ingram © Gardeners' World Magazine; 184 Bibliothèque nationale de France, Paris; 186c Torie Chugg © Gardeners' World Magazine; 186b Sarah Cuttle © Gardeners' World Magazine; 187, 188 Image from the BHL. Contributed by Missouri Botanical Garden, Peter H. Raven Library; 189a Paul Debois © Gardeners' World Magazine; 189b Jason Ingram © Gardeners' World Magazine; 191 Tim Sandall © Gardeners' World Magazine; 192a Giffany; 192c Jason Ingram © Gardeners' World Magazine; 192b Torie Chugg © Gardeners' World Magazine; 193 Noah Jordan; 194 Image from the BHL. Contributed by New York Botanical Garden, LuEsther T. Mertz Library; 195a Jason Ingram © Gardeners' World Magazine; 195c Clare Gainey / Alamy Stock Photo; 195b thrillerfillerspiller/Alamy Stock Photo; 196 James53145; 197, 198c Image from the BHL. Contributed by Missouri Botanical Garden, Peter H. Raven Library; 198a Florilegius/Alamy Stock Photo; 198b, 200 Jason Ingram © Gardeners' World Magazine; 199 Image from the BHL. Contributed by University of California Libraries; 201a Sunny25; 201c Florilegius/Alamy Stock Photo; 201b, 204, 205c Image from the BHL. Contributed by Missouri Botanical Garden, Peter H. Raven Library; 202 guentermanaus; 203 Paul Debois © Gardeners' World Magazine; 205a Bryony van der Merwe; 205b Liviu Gherman; 206a Peter Turner Photography; 206c Sarah Cuttle © Gardeners' World Magazine; 206b Nikolay Kurzenko; 207 Image from the BHL. Contributed by MBLWHOI Library; 208 Image from the BHL. Contributed by Missouri Botanical Garden, Peter H. Raven Library; 209 AlyssaRich; 211 Avalon.red/Alamy Stock Photo; 212a Wiert nieuman; 212c Sakss; 212b Tony Baggett; 213 Image from the BHL. Contributed by Missouri Botanical Garden, Peter H. Raven Library; 214, 215a Jason Ingram © Gardeners' World Magazine; 215c Nikolay Kurzenko; 215b AN NGUYEN; 216 Jason Ingram © Gardeners' World Magazine; 217 MacBen; 218a Clare Gainey/Alamy Stock Photo; 218c Nahhana; 218b, 219 Image from the BHL. Contributed by Missouri Botanical Garden, Peter H. Raven Library; 220 Cleveland Museum of Art, Cleveland; 221a Paul Debois © Gardeners' World Magazine; 221c Juta; 221b Jason Ingram © Gardeners' World Magazine; 222 KPG-Payless2; 223, 225 Image from the BHL. Contributed by New York Botanical Garden, LuEsther T. Mertz Library; 224 Florapix/Alamy Stock Photo; 226a John Richmond/Alamy Stock Photo; 226c flowerphotos/Alamy Stock Photo; 226b thrillerfillerspiller/Alamy Stock Photo; 227 Florilegius/Alamy Stock Photo; 228 Album/Alamy Stock Photo; 229a Image from the BHL. Contributed by Auckland War Memorial Museum Tāmaki Paenga Hira; 229b Image from the BHL. Contributed by Harvard University Botany Libraries; 231 Image from the BHL. Contributed by Missouri Botanical Garden, Peter H. Raven Library; 232a Taras Verkhovynets; 232c Jason Ingram © Gardeners' World Magazine; 232b Florilegius/Alamy Stock Photo; 233, 234, 235a, 235b, 236 Image from the BHL. Contributed by Missouri Botanical Garden, Peter H. Raven Library; 235c Sarah Cuttle © Gardeners' World Magazine; 237 Nyan Xing; 238a Album/Alamy Stock Photo; 238c Tom Meaker; 238b Andrew Darrington/Alamy Stock Photo; 239 The History Collection/Alamy Stock Photo; 240 Image from the BHL. Contributed by Missouri Botanical Garden, Peter H. Raven Library; 241a Barbarajo; 241c weruu; 241b, 242 Sarah Cuttle © Gardeners' World Magazine; 243 knelson20; 244 Zdelnik Alexandr; 245a Image from the BHL. Contributed by Missouri Botanical Garden, Peter H. Raven Library; 245c John Richmond/Alamy Stock Photo; 245b Image from the BHL. Contributed by New York Botanical Garden, LuEsther T. Mertz Library; 246a RM Floral/Alamy Stock Photo; 246c Donna Bollenbach; 246b Jason Ingram © Gardeners' World Magazine; 247 Album/Alamy Stock Photo; 248 Rudi Sebastian/Alamy Stock Photo; 249 Joe Kuis; Front and Back Cover Images from the BHL. Contributed by Missouri Botanical Garden, Peter H. Raven Library

1

BBC Books, an imprint of Ebury Publishing
Penguin Random House UK
One Embassy Gardens, 8 Viaduct Gdns,
Nine Elms, London SW11 7BW

BBC Books is part of the Penguin Random House group of companies whose addresses can be found at global.penguinrandomhouse.com

First published by BBC Books in 2024

www.penguin.co.uk

A CIP catalogue record for this book is available from the British Library

ISBN 9781785948954

Commissioning Editor: Phoebe Lindsley
Writer: Tamsin Hope-Thomson
Design: Peter Dawson, Ronja Ronning, www.gradedesign.com
Production: Phil Spencer

BBC Gardeners' World Magazine: Kevin Smith, Tamsin Hope Thomson, Sarah Edwards
Compiled by BBC Gardeners' World Magazine

Printed and bound in Estonia by Print Best

The authorised representative in the EEA is Penguin Random House Ireland, Morrison Chambers, 32 Nassau Street, Dublin D02 YH68.